UNIVERSITY SCHOOL LIBRARY
UNIVERSITY OF WYOMING

THE ELECTROMAGNETIC SPECTRUM
KEY TO THE UNIVERSE

EXPLORING OUR UNIVERSE

THE ELECTROMAGNETIC SPECTRUM

KEY TO THE UNIVERSE

Franklyn M. Branley

ILLUSTRATED BY Leonard D. Dank

THOMAS Y. CROWELL   NEW YORK

Illustration on page 111 © Association of Universities for Research in Astronomy, Inc., Sacramento Peak Observatory. From "Ultraviolet Astronomy" by Leo Goldberg, copyright © 1969 by Scientific American, Inc. All rights reserved.

Text copyright © 1979 by Franklyn M. Branley
Illustrations copyright © 1979 by Leonard D. Dank
All rights reserved. Printed in the United States of America. No part of this book may be used or reproduced in any manner whatsoever without written permission except in the case of brief quotations embodied in critical articles and reviews. For information address Thomas Y. Crowell, 10 East 53rd. Street, New York, N.Y. 10022. Published simultaneously in Canada by Fitzhenry & Whiteside Limited, Toronto.
Designed by Trish Parcell

LIBRARY OF CONGRESS CATALOGING IN PUBLICATION DATA
Branley, Franklyn Mansfield, 1915-   The electromagnetic spectrum.
SUMMARY: An explanation of the nature of the electromagnetic spectrum, the scientific investigations that led to our understanding of it, and its role as a tool of science.
1. Electromagnetic waves—Juvenile literature.
[1. Electromagnetic waves] I. Title.
QC661.B663   1979   537   77-26591
ISBN 0-690-03868-2   ISBN 0-690-03869-0 lib. bdg.

First Edition

BY THE AUTHOR

THE CHRISTMAS SKY
COLOR: FROM RAINBOWS TO LASERS
ENERGY FOR THE TWENTY-FIRST CENTURY
EXPERIMENTS IN THE PRINCIPLES OF SPACE TRAVEL
MAN IN SPACE TO THE MOON
THE MYSTERY OF STONEHENGE
PIECES OF ANOTHER WORLD: THE STORY OF MOON ROCKS
SOLAR ENERGY

Exploring Our Universe

THE NINE PLANETS
THE MOON: EARTH'S NATURAL SATELLITE
MARS: PLANET NUMBER FOUR
THE SUN: STAR NUMBER ONE
THE EARTH: PLANET NUMBER THREE
THE MILKY WAY: GALAXY NUMBER ONE
COMETS, METEOROIDS, AND ASTEROIDS:
   MAVERICKS OF THE SOLAR SYSTEM
BLACK HOLES, WHITE DWARFS, AND SUPERSTARS
THE ELECTROMAGNETIC SPECTRUM: KEY TO THE UNIVERSE

Contents

1. A Keyboard of Energy 1
2. Matter and Energy 7
3. Light 16
4. Light and Electromagnetism 35
5. Electromagnetic Waves 43
6. Max Planck: Packets of Energy 48
7. Albert Einstein: The Photon Theory 56
8. Atoms and the Spectra of Gases 63
9. The Visible Spectrum 71
10. The Radio Spectrum 84
11. The Infrared Spectrum 101
12. Ultraviolet, X Rays, and Gamma Radiation 110

   For Further Reading 122
   Index 123

THE ELECTROMAGNETIC SPECTRUM
KEY TO THE UNIVERSE

1 A Keyboard of Energy

Every day of your life you are affected by the electromagnetic spectrum. It is your main contact with the world. For example, you see things because of light energy. Light is part of the electromagnetic spectrum, the part that makes possible all that you learn by seeing.

Radio signals are electromagnetic waves that carry all sorts of information—from broadcasting stations and also from stars and galaxies. Television signals are also electromagnetic waves. The space around the earth is full of them. Some reach your set directly. Other signals are relayed by satellites and so give you an on-the-spot and at-the-moment view of events occurring on the other side of the world.

Another part of the electromagnetic spectrum is made up of infrared waves. Infrared radiation is energy which comes mainly from the Sun: the energy that keeps you warm and stimulates plant growth, and so is essential to life itself. When you are outdoors on a bright day ultraviolet radiation falls on you. That's another part of the electromagnetic spectrum.

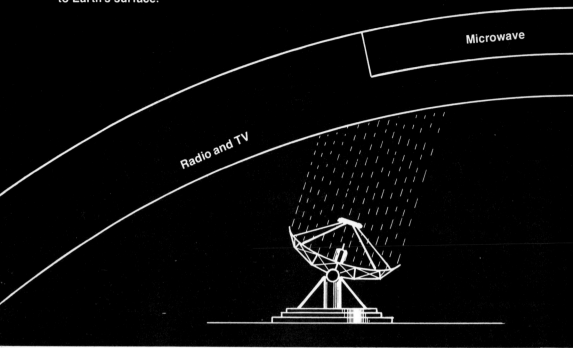

In space there is radiation from long radio waves to very short gamma rays. Only radio and light waves can penetrate through the atmosphere to Earth's surface.

You're not aware of them, but you are constantly exposed to X rays, which are also electromagnetic. They are produced in certain stars, including the Sun. Most are filtered out by the atmosphere but some penetrate to the surface of the earth, especially to high mountaintops, where the air is thin.

The electromagnetic spectrum is a chart of energy. You might compare it with a chart of the keyboard of a piano. The keys on the left side produce low notes; these sounds have long wavelengths. As in ocean waves, there is a long distance from the top of one of these sound waves to the top of the next. When you move to the right, striking notes as you go, the sounds become higher—these sound waves have shorter

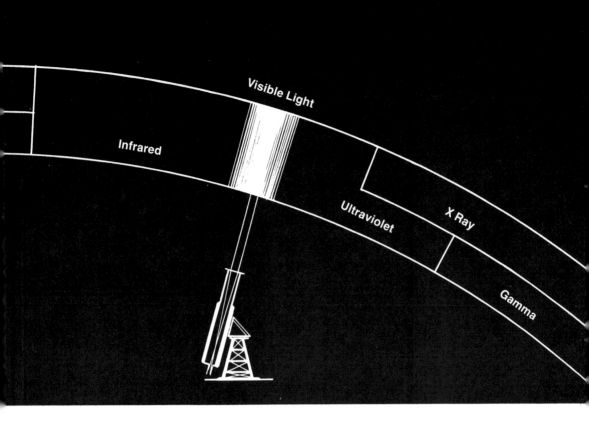

wavelengths. In some ways the information represented on the electromagnetic chart is similar; toward one end are the long radio waves, light waves are shorter, and X rays, toward the opposite end, are much shorter than light waves.

That's about the only way that the two charts are similar. Sound is not part of the electromagnetic spectrum. The comparison is used here only because both sound waves and electromagnetic waves vary in length, and because the arrangement of the keyboard of the piano helps you to understand the chart of the electromagnetic spectrum. Sound differs from light and other parts of the spectrum in many ways. For example, sound travels about 340 meters in a second;

light travels 300 000 000 meters in a second, and so do all other parts of the spectrum—radio waves, television signals, X rays.

LIGHT LEADS TO WONDER

Light is the kind of electromagnetic energy that scientists learned about first. You'd expect this to be so, for light is the only part the spectrum we are sensitive to. We have eyes that respond to those wavelengths that produce light. But there are no parts of our bodies that are conscious of radio waves, X rays, or gamma radiation.

Finding an explanation for light and determining how it gets from one place to another were tremendous puzzles for scientists. Over the course of several centuries dozens of scientists studied light. They discovered that it travels very fast, that it travels in waves, and that red light is different from violet light because of differences in wavelength. As knowledge of light increased, some investigators wondered if there were other forms of energy that traveled in waves— energy that was not apparent, but nevertheless might exist.

Everlasting curiosity led scientists beyond visible light, into regions of longer and longer wavelengths—longer than light; and also into regions where the wavelengths were shorter than those of light. A pattern began to emerge; the information that was being gathered gradually became organized. Red light was known to be made of long waves, and violet light of shorter waves; infrared was beyond the red, while ultraviolet was beyond the violet. The various wavelengths could be separated into groups, one blending into another, and a chart representing them could be constructed. Refinements and enhancements of the chart have been made for decades, and they continue to be made today.

THE MEANING OF "ELECTROMAGNETIC"

The chart of energy has been called the electromagnetic spectrum. You know about the color spectrum; it is the rainbow of colors produced when sunlight is dispersed by a prism into its various parts—red, orange, yellow, green, blue, and violet. The spectrum of energy is an extension of the color spectrum—the shorter waves being those beyond violet, and the longer waves those beyond red. As you can see from the chart on pages 2–3, the color section is a very small part of the total energy spectrum.

The energy is called "electromagnetic" because it is related to both electricity and magnetism. The two exist together—you cannot have one without the other. When electricity flows there is a magnetic field around the conductor. When you have magnetism you can produce electricity by moving a conductor through the magnetic field. An electric generator turns a coil of wire in a magnetic field; this causes an electric current to flow through the coil. Magnetism can affect an electric current, and an electric current affects magnetism.

Radio waves can alter an electric current, as they do in a radio or television set; light waves or infrared waves can turn electricity on and off to operate doors and alarm systems. Theoretically one part of the spectrum can be changed into any other part—just by changing the energy level. All the parts are related—new facts about one part lead to knowledge about other parts.

EXPLORATION CONTINUES

During the twentieth century a prime purpose of scientific investigation has been to explore the energy of the universe.

This has been done with microscopes, optical and radio telescopes, spectroscopes, X-ray machines, gamma-ray and X-ray satellites; with thousands of different kinds of instruments and devices. They are helping scientists to fill in the electromagnetic chart with information—to learn more about all parts of it and to discover how we can use the energy to understand ourselves and the whole universe better.

Much of our knowledge about the parts of the chart was obtained during the past few decades, and just about all of it within the past century. Our way of life depends upon that knowledge. It has given us awareness of everything around us, from the submicroscopic world to the world of stars and galaxies. This idea was expressed by R. Buckminster Fuller, an American scientist and engineer: "Up to the twentieth century 'reality' was everything humans could touch, smell, see and hear. Since the initial publication of the chart of the electromagnetic spectrum humans have learned that what they can touch, smell, see and hear is less than one-millionth of reality. Ninety-nine percent of all that is going to affect our tomorrows is being developed by humans using instruments and working in ranges of reality that are nonhumanly sensible."

2 Matter and Energy

When the universe is reduced to its fundamentals, there are only two basic parts—matter and energy.

Matter is mass, so all those things that have mass—stars, planets, galaxies, iron and oxygen, gold and lead, you and I—make up the matter of the universe. Energy is more difficult to define. At the start we shall describe energy by what it does. It travels very fast, some 300 000 kilometers a second (186,000 miles per second), in waves that can be reflected, refracted (that is, bent), or controlled in other ways. This is radiant energy—energy that radiates into space. It is the energy that is charted in the electromagnetic spectrum. We'll explore its many parts and find ways in which each part differs from the others. But first let's consider ways in which they are the same.

Information about the spectrum of energy was obtained slowly. Because early researchers had no way of knowing there were such things as radio waves and X rays, the information came from the study of visible light. Much was known about light before anyone realized that other, different forms of energy existed, and once these forms were discovered people still did not realize that they were related. This was to be expected, for the existence of radio waves was not even suspected until the beginning of the nineteenth century, and they were not discovered until near the end of the century. Yet even the ancients knew that light existed. They could not explain light, but they knew that it was related in some way to the Sun.

Light travels in waves, and so do all other parts of the spectrum. Also, light travels at a speed of 3×10^{10} cm/sec (often given as 3×10^5 km/sec); and so do all parts of the spectrum. (This is called powers-of-10 notation: 30 000 000 000 is written 3×10^{10}—ten zeros after the 3. Values less than 1, such as .00003, are written 3×10^{-5}—that is five places to the right of the decimal.)

Whether radiant energy is a radio wave, visible light, or a gamma ray depends upon the wavelength of the energy or its frequency. You might compare a wave of radiant energy to a wave in the sea. Both of the waves have high points, called crests, and low points, called troughs. The distance from one crest to the next is the wavelength.

Long-wave radio radiation may have a wavelength of 1 000 meters or more; there may be a thousand meters between two crests of a radio wave. Wavelength gets shorter and shorter as one moves toward gamma rays, which have the shortest wavelength. Gamma radiation may have a wavelength of less than .000 000 001 (one billionth) of a meter—called a nanometer.

Velocity (or speed) is related to wavelength and frequency of the waves. Frequency is the number of waves passing a

On the left is long-wave radiant energy; a shorter wave is on the right. The waves are three dimensional—they vibrate in all planes around the center line. Only the vertical and horizontal planes are shown.

Wavelength

10   THE ELECTROMAGNETIC SPECTRUM

given point in a second. It is usually given in cycles per second. For example, red light can be expressed as radiation that has a frequency of 4×10^{14} cm/sec.

Wavelength times frequency gives velocity (velocity = wavelength × frequency), so we can find frequency easily when wavelength is known. We know that the velocity of radiation is always essentially 3×10^{10} cm/sec. Suppose the wavelength of the radiation is .001 centimeter. We can find its frequency by using the equation:

$$3 \times 10^{10} = .001 \times \text{frequency}$$
$$\text{Frequency} = \frac{3 \times 10^{10}}{.001} = \frac{30\ 000\ 000\ 000}{.001}$$
$$\text{Frequency} = 30\ 000\ 000\ 000 \text{ cm/sec}$$

If we know the frequency of the energy, we can find the wavelength in a similar manner. For example, suppose the frequency is 3×10^5 cm/sec; then:

$$\text{Velocity} = \text{wavelength} \times \text{frequency}$$
$$3 \times 10^{10} = \text{wavelength} \times (3 \times 10^5)$$
$$\text{Wavelength} = \frac{30\ 000\ 000\ 000}{300\ 000}$$
$$\text{Wavelength} = 100\ 000 \text{ cm}$$

From the equation it is apparent that when wavelength decreases, frequency increases. The frequency of the long-wavelength radio waves we were discussing earlier is about 300 000 cm/sec, while the frequency of the short-wavelength gamma radiation is some 30 000 000 000 000 000 000 (3×10^{19}) cm/sec.

The greater the frequency of the radiation, the greater the energy involved. It takes more energy to produce gamma radiation than to produce radio waves.

TYPE OF RADIATION	ENERGY RELATIVE TO THE ENERGY OF LIGHT
Radio waves	2×10^{-9}
Infrared	5×10^{-2}
Light	1
Ultraviolet	50
X rays	5×10^{3}
Gamma rays	5×10^{4}

Relationships between frequency and wavelength, as well as other apsects of radiation, are shown on pages 2 and 3 in a representation of the entire electromagnetic spectrum. Long radio waves appear at the left and short gamma waves at the right.

When early researchers were developing this plan for arranging the spectrum, some believed that the shortest waves should be shown at the left and the longest at the right. Reading left to right, there would be progression from short waves to long. The arrangement seemed logical to them. Others argued that it was just as logical to have the longer wavelengths displayed at the left and to develop toward shorter waves on the right. Besides, these people argued, that's the way a piano is arranged. Low notes (long waves) are on the left, and the high notes (short waves) are on the right. The "piano" argument won out, and so long waves are usually shown on the left and short waves on the right—although either arrangement is correct.

Notice that there is no cut-off at the extremes; there are no ends to the spectrum. Our knowledge of the spectrum is limited. We cannot say it begins at one point and ends at another. Theoretically radiation, especially of the shorter

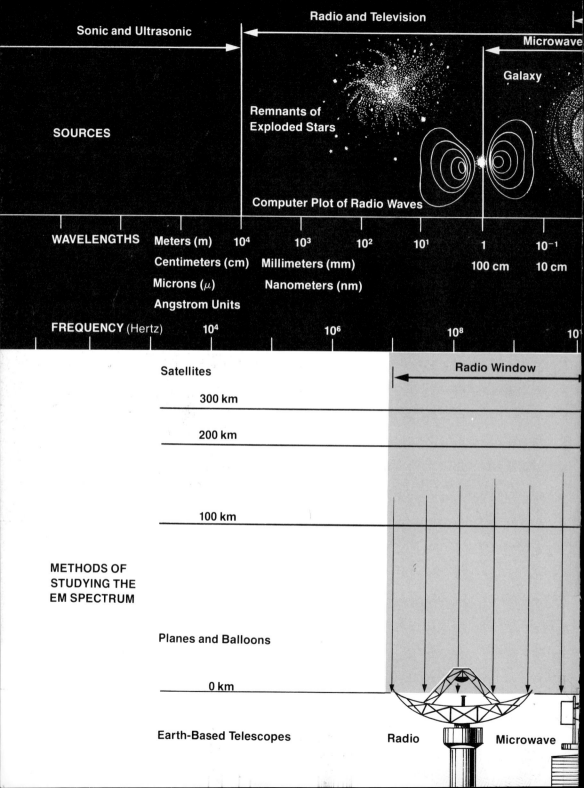

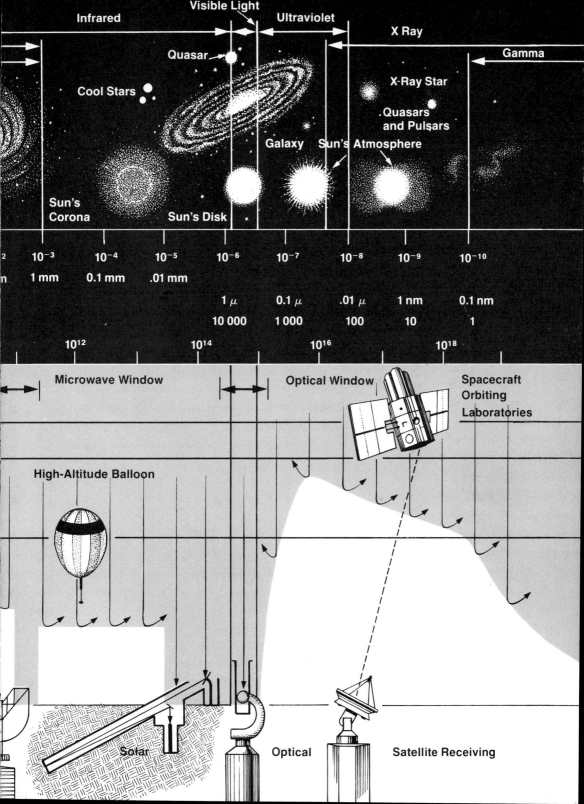

wavelengths, has no end, although it is hard to conceive how wavelengths far beyond gamma radiation could be produced because of the tremendous energy needed. Notice also that areas overlap; microwaves overlap radio waves, ultraviolet blends into X rays, X rays into gamma rays.

Wavelengths of the various kinds of radiation are given in the bands. You'll notice that we use different terms—such as meters, microns, and nanometers. Different units are needed because the range is very wide and a single unit becomes awkward. Suppose, for example, that we used only centimeters. The range from long radio waves to gamma radiation is from about 2 000 000 centimeters to .000 000 000 05 (five-hundred-billionths of a) centimeter. Such large and small numbers can be handled more efficiently by using powers of 10, as we have been doing. Then 2 000 000 becomes 2×10^6, and 0.000 000 000 05 becomes 5×10^{-11}.

Even so, the measurements have more meaning when they are expressed in units more appropriate to the dimensions. Kilometers are used for the longest waves. As the waves become shorter, meters and centimeters are used. For even smaller waves we use microns and nanometers. You may also see angstroms used for very short waves; however, the nanometer is the accepted unit. A micron is one ten-thousandth, or 10^{-4}, of a centimeter. Or, in terms of a meter, a micron is one millionth—10^{-6}. One ten-billionth of a meter is called an angstrom. Ten angstroms equal one nanometer. The prefix *nano-* signifies 10^{-9}, or one billionth.

You'll notice that frequency of the various radiations ranges from 60 vibrations a second to 600 000 000 000 000 000 000 (6×10^{20}) vibrations a second.

The low-energy vibrations are produced by an electric generator. The alternating current used in the United States is 60-cycle AC, with a wavelength of some 5 000 kilometers. High-energy gamma radiation is so energetic it can be produced only in certain superhot stars or by large clusters of stars such as those in the central region of our galaxy.

Our knowledge of the electromagnetic spectrum is relatively new. All of the radiation is invisible, requiring special equipment to identify it. Light is made of those particular wavelengths to which our own special equipment, our eyes, are sensitive. No equipment had to be devised for us to be aware of its presence. Therefore, many early scientific adventures concerned light. Early scholars were curious about it: what was light, how was it generated, and how did it travel from place to place?

3 Light

You would have a hard time explaining what light is to someone who knew nothing about it. And understandably so, for after centuries of investigation not even scientists have found a completely acceptable definition.

Two thousand years ago there was disagreement among philosophers as to the nature of light. Many Greek scholars believed that light was made up of particles: we saw things because somehow the objects threw off bits of matter. These particles traveled through space and entered our eyes, and so the object could be seen. Other Greek scholars, Aristotle among them, could not accept this idea. They believed that light resulted from some kind of action (just what kind they did not know) that took place in the space between the object and the viewer. The space itself was transparent.

Better explanations than these were not achieved until the seventeenth century. At that time numerous scientists were probing the mysteries of light. Their discoveries brought us a lot closer to understanding what light is and how it behaves.

THE SPEED OF LIGHT

In the early 1600s, Galileo was interested in finding the speed of light. He wondered if any time was required for light to go from place to place. Galileo, unlike scientists who had lived before him, did not just reason out answers; he experimented. He attempted to measure the time needed for a light signal to travel from one hill to another and then return. He tried the experiment but it just would not work. You can see why when you consider that in the days of Galileo the best light sources were oil lanterns or candles. The hills had to be close together—otherwise the light could not be seen. In those days there were no devices for measuring even seconds of time, let alone the small fraction of a second that it takes light to travel from one hilltop to its neighbor. So, although his idea was correct, the equipment available to Galileo was not equal to the task.

Later in the century, in 1675, The Danish astronomer Ole Roemer (1644–1710) obtained a remarkably accurate idea of the speed of light by following an enirely different procedure. He observed Io, then believed to be the innermost satellite of Jupiter. He noted the time when the satellite moved behind Jupiter (when it was eclipsed by Jupiter), how long it remained there, and when it was seen again. The satellite was found to have a constant period of revolution around Jupiter of 42½ hours.

The period was so regular that Roemer could predict accurately one year ahead the times when eclipses of Io would occur. But his predictions for six months ahead were inaccurate. Earth moves much faster in its orbit than does Jupiter. Therefore Earth catches up to Jupiter and passes the planet;

it moves alternately toward the planet and away from it. When Earth was moving away from Jupiter, the satellite seemed to slow down. It moved into Jupiter's shadow later and later than the predicted time. As Earth moved toward Jupiter, the satellite appeared to speed up; the intervals between eclipses became shorter and shorter than predicted. When the distance between Earth and Jupiter was not changing (Earth was not moving toward or away from Jupiter) the interval between eclipses once more became 42½ hours.

When Earth was farthest from Jupiter, the interval between eclipses of Io was slightly more than 42 hours 46 minutes—about 1,000 seconds longer than when Earth was closest to Jupiter. Roemer said this was because the light had to travel farther; it had to go across the entire diameter of Earth's orbit. In Roemer's time it was believed that the Sun was about 150 million kilometers away, so the diameter of the orbit would be twice that, or 300 million kilometers. Since the light from Io took 1,000 seconds to travel 300 million kilometers, the light must travel 300 thousand kilometers in one second:

$$\text{Velocity} = \frac{300\,000\,000}{1\,000} = 300\,000$$

Incredible, many scientists said, in attacking Roemer's conclusion. It was ridiculous to think that anything could travel so fast. It was easier to believe that light was instantaneous—that it happened all at once.

About fifty years later the English astronomer James Bradley (1692–1762) discovered another way to measure the speed of light, one that also used astronomical observations. He discovered that the positions of stars shifted slightly as Earth moved in its orbit around the Sun. The shift is called the aberration of starlight.

You can get an idea of why it happens if you consider an umbrella carried as a person walks in the rain. Suppose the rain is coming straight down. When the person is standing still, he holds the umbrella directly over his head. Now suppose he starts walking. To protect himself from the rain, he must tilt the umbrella slightly in the direction of movement. The faster he walks, the greater the tilt.

Bradley knew that since Earth moved, telescopes must be tilted to observe the stars, just as the umbrella must be tilted toward the rain. He found that the tilt—the angle of aberration—was 20.5 seconds of arc, or about ⅓ degree. He also knew that Earth moves about 18 miles a second in its journey around the Sun. From these two bits of information, Bradley computed that starlight must be traveling some 186 000 miles a second. (Had he been using the metric system, Bradley would have written it 3×10^5 km/sec.)

This is how the reasoning goes. Suppose a telescope were 3×10^5 kilometers long. The telescope is being carried forward at 29 km/sec. Light entering the telescope moves in a straight line down the telescope tube. If the telescope were not tilted, the light would soon strike the side of the instrument. But since it is tilted, the light reaches the bottom of the tube exactly as the telescope completes a distance of 29 kilometers. Therefore the light traveled the length of the tube (3×10^5 kilometers) in one second.

More than a hundred years were to go by before these astronomical methods for measuring the speed of light would be augmented by methods that gave more precise results.

About the middle of the nineteenth century a Frenchman, Armand Hippolyte-Louis Fizeau (1819–1896), measured the speed of light using an idea similar to the one Galileo had

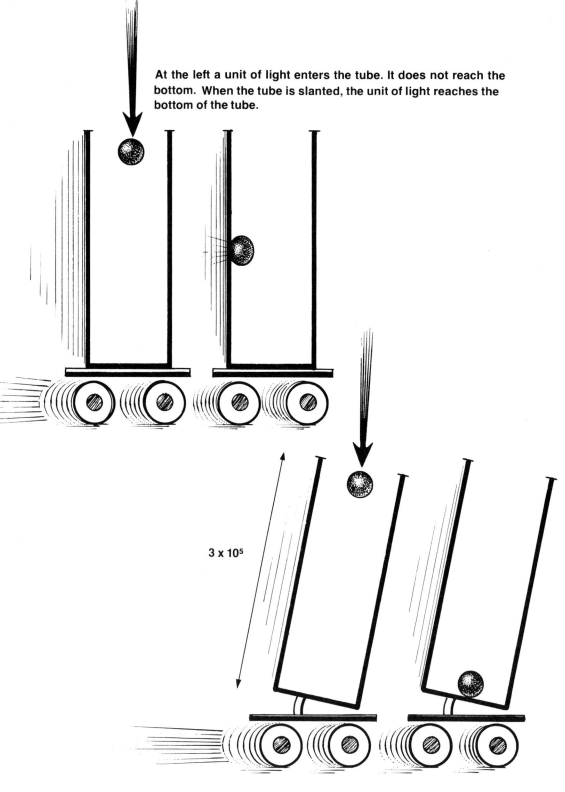

tried some two hundred years earlier. He measured the time it took for a pulse of light to travel a given distance. Fizeau made light pulses by turning a toothed wheel through a beam of light. The light went to a distant mirror and was returned. If the light had passed through a notch in the wheel, it was bright when it returned. If it had struck a tooth, the returned light was dim. By observing the light and varying the speed of the wheel, Fizeau could determine the time needed for the light to travel to the mirror and back. He knew the distance and so could figure out how fast the light had to travel to cover that distance in the measured time. According to his calculations, light traveled 313 300 km/sec. Later investigators who had more accurate equipment found that the figure was much too high, but the procedure that Fizeau had invented was correct. In fact, after being improved upon by other experimenters, it provided for a time the most reliable measurement of the speed of light.

In 1873, Albert A. Michelson, an American astronomer who was then twenty-one years old, found the speed of light to be 299 853 km/sec, using a system similar to Fizeau's. (The most accurate modern figure for the speed of light is 299 793 km/sec.) In 1923, after he had developed better equipment, Michelson tried again. In place of the toothed wheel he used a rotating mirror, placed on Mt. Wilson in California. Exactly 35.4438 kilometers away he placed another mirror atop Mt. San Antonio. The distance was exact to one centimeter in 14 kilometers. After three years of experimenting with the apparatus Michelson announced that the speed of light was 299 796 km/sec. This is very close to the figure now accepted. While the velocity should be known as precisely as possible, such extreme accuracy is not needed

22   THE ELECTROMAGNETIC SPECTRUM

in most calculations. For the speed of light the figure 3×10^{10} cm/sec is used most often—or, in the customary system, 186,000 miles per second.

The reflection of radar signals from the moon and planets, achievements realized since the advent of the space age, have corroborated Michelson's findings. So also have the data obtained by measuring the time required to send laser pulses to the moon, bounce them from reflectors left there by Apollo astronauts, and pick them up with Earth-based receivers.

LIGHT WAVES OR LIGHT PARTICLES

Roemer's revelation that the speed of light was fantastically high upset those persons who believed that light was made of particles. It was inconceivable that any mass could go so fast. Nevertheless Isaac Newton persisted in his belief that light consisted of streams of particles given off by whatever was producing the light.

There were a few who disagreed with the mighty Newton. Prominent among them was the Dutch scientist Christian Huygens (1629–1695). He developed a theory that held light to be a wave phenomenon that traveled through the ether. At that time it was believed that sound waves, traveling through air, and water waves, traveling through water, were transmitted when one particle hit against another. Light, Haygens argued, traveled in a similar fashion, from "ether molecule to ether molecule."

Ether was considered to be a substance that filled all space. (The name comes from a Greek word meaning "the upper air," and has nothing to do with the anesthetic.) For over two hundred years it was believed that this ether supported the

Light 23

passage of light. Its presence was not disproved until the beginning of this century, when Michelson, who had measured the speed of light, and E. W. Morley, his colleague, performed an experiment that made it possible to conclude there was no such substance.

Huygens believed that light waves were like sound waves; they vibrated in the direction of travel. The idea was not new. Leonardo da Vinci (1452–1519) had suggested the possibility after he had studied water waves and sound waves. Later it was found that light waves vibrate at right angles to the direction of travel. When the waves passed from the either into water, the light was bent. Huygens said this happened because the light was slowed down.

Not so, said Newton. The light was bent, he agreed, but it did not slow down. It speeded up, he said, because the water attracted the light particles.

At that time no way had been found to determine whether the light speeded up or slowed down. This argument of the seventeenth century was not settled until about the middle of the nineteenth century, when the French scientist Jean B.-L. Foucault (1819–1868) was able to measure the speed of light in water. He found it to be considerably slower than the speed of light in air. This experiment was a big factor in settling the age-old controversy about the nature of light. Apparently light was made of waves that traveled through the either.

LIGHT AS RADIANT ENERGY

Even today we probably do not understand the true nature of light. We do know that it is a form of radiant energy that travels through space and needs no "ether," or anything else, to carry it. However, when light falls upon certain substances —the selenium in a light meter, for example—it behaves as though it is made of particles. A particle containing a certain amount of energy will cause a certain amount of electricity to be generated. The electricity causes a pointer to indicate the "quantity" of light falling upon the meter.

Sometimes, it seems, light behaves as a wave. At other times it appears to be a particle. Some people get around the dilemma by saying that light is a "wavicle." Later on we'll come back to this problem.

INTERFERENCE OF LIGHT

Because Newton was a giant in his time, anything that he said, or supported, was believed to be true by most of his followers. Therefore, even long after he had died, in 1727, his particle theory of light was accepted widely.

However, in 1803, Thomas Young (1773-1829), an Englishman who was not only a famous scientist but a teacher and physician as well, published a paper that completely refuted Newton's conception of light particles—or so it seemed. So strong was the opposition of the Newtonites to Young's findings that he put them aside for a while, allowing time for the furor to quiet down.

Young directed a beam of light through a narrow slit. Beyond the slit was a barrier with two other slits in it, each at an angle to the first. If light was made of particles that traveled in straight lines, Young reasoned, two bright bars of light and nothing else should appear on a screen placed beyond the second barrier.

But there were several vertical bars of light—in fact, a series of light and dark bars. Young said that this was because

Light travels in waves. It passes through a slit (bottom), and then passes through two slits. If it were a beam, it could not do this. The waves cancel and augment each other, producing a band having alternate light and dark areas (top).

light travels in waves. A wave radiates from the first slit. When the wave hits the slits in the second barrier, two new waves are created. When the waves support each other (one crest being added to the other) there is a bar of light. When the waves are out of step (a crest and a trough combined) the waves cancel each other, and there is a bar of darkness.

When the light that Young used was of one color (monochromatic) the light and dark bands were sharp at the edges. But when white light, which is a combination of all colors, was used, there were fringes of color on either side of the bright bars. Young reasoned that the different colors were due to different wavelengths. He had succeeded in showing the interference of light, a phenomenon possible only because light travels in waves.

NEWTON'S VISIBLE SPECTRUM

Fifty years before Young's experiment, Newton had noticed the same color fringes that appeared in Young's experiment, and he had explained them. Among his other accomplishments, Newton was an astronomer. He and other astronomers were disturbed by the images seen in telescopes. These images were not clear, and they had fringes of color around them. Newton solved this problem and, in so doing, discovered that ordinary white light is composed of light of many different colors. These colors were later found to be due to different wavelengths.

Newton placed a glass prism in the path of a beam of sunlight that traveled through a narrow slit. An array of colors was produced: red, orange, yellow, green, blue, and violet. He called the spread of colors a spectrum. (The word "spectrum" means "image.") A spectrum is a series of im-

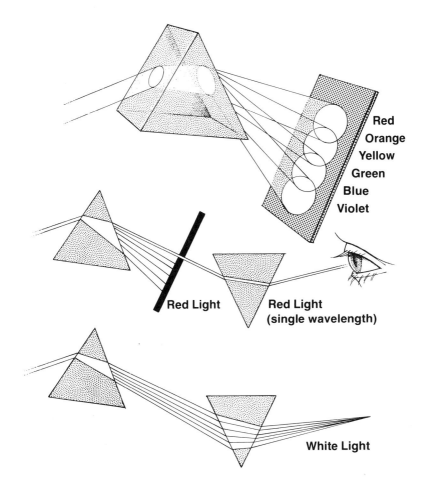

White light is composed of wavelengths that produce a complete spectrum of colors when they are separated (top). Red light is "pure" because it is unchanged by a second prism. The entire color spectrum is recombined into white light by a second prism (bottom).

ages of the slit through which the light travels. The images are so many, and they are so close together, that an observer cannot separate one from another; the images blend together.

Newton then placed a barrier with a slit in it in the red part of the spectrum. He directed the red light through a second prism. When the light was observed, no change had occurred; it was still red. The light was already in its simplest form—it was a single wavelength, as Thomas Young was to discover later on.

However, when Newton passed the entire spectrum of colors through a second prism, white light was produced. Newton suspected that the results he observed—the spectrum of colors, the red light, the recombining to produce white light—occurred because different colors were due to different wavelengths. He could not prove this, however. Also, Newton was committed to the particle theory of light and so probably did not actively pursue the "wave explanation."

We know now that long-wave red light, with a wavelength of about 7.5×10^{-5} centimeters (750 nanometers), is bent the least in passing through a prism. Short-wave violet light, with a wavelength of about 3.5×10^{-5} centimeters (350 nanometers), is bent the most.

As a result of his work with prisms and color fringes, Newton the astronomer invented the reflecting telescope. This instrument uses mirrors rather than lenses to collect light. Mirrors produce no color fringes.

THE DIFFRACTION GRATING

Young's experiment, which showed the wave nature of light, consisted of passing a narrow beam of light through two slits. The plate with two slits in it was the forerunner of the diffraction grating that is used today in many light-analyzing instruments. It is a plate that may have six thousand or more slits (or lines) in a single centimeter.

To make the grating, a remarkably precise machine scratches lines on a piece of fine glass, or metal if the grating is to reflect light rather than pass it. If light shining on the grating were to fall in a straight line on a screen placed behind it, an observer would see the light on the screen; it would not be changed in any way. However, if the screen were placed at an angle to the grating, one would see an array of colors on the screen. The viewer would "see" the various wavelengths of which the light is composed. Knowing the distance between the slits and the angle at which the light leaves the grating, it is possible to compute the wavelength of the light. The difference in length between two adjacent light beams turns out to be the wavelength of the light.

When light of one single wavelength passes through the grating, the light that leaves the grating is all diffracted, that is, spread, at the same angle. The light that emerges from the grating is the same as the light that enters it. However, most light is not made of a single wavelength. While light may appear to be one color, there will most likely be several wavelengths combined together.

A rather pure color can be made quite simply, however. If you sprinkle salt (sodium chloride) in a gas flame, the flame becomes yellow. The color is produced by the sodium. If this light is passed through a diffraction grating only two lines will be observed. Both will be in the yellow region of the spectrum. Sodium produces light which consists of only two wavelengths, and these are close together.

On the other hand, if a neon tube (which appears to be one color) is observed through a grating, a series of many bright lines will be seen. If the tube contains mercury vapor, or argon, or some other gas, different lines will be produced. This is because each of the wavelengths that compose the color produced by the gas forms an image of the slit through

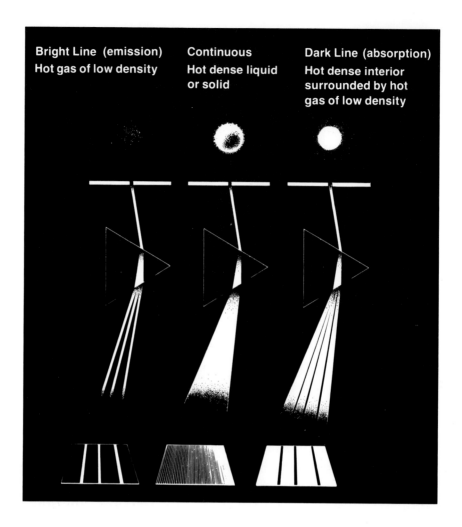

which the light passes. And the images will always be at the angles determined by the particular wavelengths of the various parts of the light.

Such arrays are called bright-line spectra. They are produced by hot gases of low density—gases which are not packed together tightly. In essence, the theory holds that when a gas is heated, units of energy called quanta or photons are picked up by electrons, the fast-moving negative charges that revolve around the nucleus of an atom. Since the electrons have become more energetic, they move outward from the center of the atom; the atom is excited. An electron will jump back to its original location because it is normal for the atom to be at rest—to be unexcited. When the electron falls back, a photon, having a certain amount of energy, or wavelength, is given off. When the gas has low density, the photons are separate and isolated; they are discrete. The lines they produce are also discrete. These are the sharp lines one sees in a bright-line spectrum.

When light is produced by hot liquids, or hot solids, or gases that are densely packed, the spectrum that is produced is continuous. The bright lines of color blend imperceptibly —red to orange to yellow, green, blue, and violet. The band is composed of a multitude of images of the slit through which the light passes. The atoms in the source of light are packed together tightly, and the photons they give off cannot be seen discretely. The light produced by any one photon blends with the light produced by photons from other atoms.

When a hot, dense interior is surrounded by hot gases of low density, the array that one sees is called a dark-line spectrum. This is a series of dark lines against a continuous spectrum, or band of colors. Wavelengths produced in the interior are intercepted by the low-density atoms in the outer envelope. Therefore, some of the energy is absorbed by these atoms. Since the energy at particular wavelengths has been diminished, the regions of the spectrum that they occupy are "cooler" than the background. They appear darker. There are many examples of the manner in which these dark-

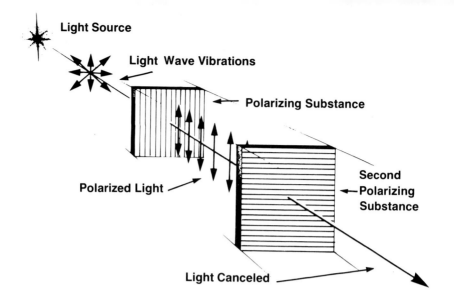

Light and other forms of radiant energy vibrate in all planes (dark arrows at left). A polarizer permits only those waves in certain planes to pass. It cancels waves in other planes.

line spectra, as well as the bright-line spectra and continuous displays, provide researchers with information.

POLARIZATION OF LIGHT

In 1802 the English scientist Thomas Young, stimulated other scientists in their study of light and especially in their efforts to determine its wave nature. A few years later a Frenchman named Augustin Jean Fresnel (1788–1827) was to devise a theory of wave motion. Experiments carried out around 1815 enabled Fresnel to set up a theory based upon mathematics—equations that could prove the wave nature of light. One of the experiments originated by Young and improved upon by Fresnel concerned the angles in which light waves vibrate: the polarization of light.

Sunglasses made of Polaroid, a special plastic material, reduce glare. When sunlight is reflected from water, for example, the light can become blinding. However, the situation can be corrected because the reflection process causes the light to become partially polarized: the light that enters your eye is vibrating in a single plane rather than in several planes. When this light goes through Polaroid glasses only a certain portion reaches your eye. Glare is removed, and the water can be seen comfortably. Often you can even see through the water to the bottom.

When sunlight passes through the atmosphere, air molecules scatter the light and also cause it to become partially polarized. Using a single Polaroid lens, look into the sky at a right angle from the Sun. Rotate the lens and you will see a change in brightness. The Polaroid filter blocks out certain waves.

When Fresnel experimented with polarization he did not have Polaroid sheet to work with—it was invented in 1932. But he did have tourmaline, a mineral that has transparent crystals that are arranged in strips, much like the slats in a Venetian blind. When Fresnel looked at light through a thin layer of tourmaline, the light was unchanged. When he looked through two thin layers, the light changed in brightness as he turned one of the sheets. The "slats" closed off the openings in the first "blind." You can do the same thing with Polaroid sunglasses—you'll need two pairs. Hold them one in front of the other and look at a light. Turn one pair. As you do, the light will get dimmer until it is cut off almost completely. This is because Polaroid contains a chemical called sulphate of iodoquinine. During manufacture, the chemical forms a mass of tiny needlelike crystals which are lined up in a single direction. Each single crystal behaves like a thin layer of tourmaline.

Fresnel's observations convinced him that light travels in waves—and that also, unlike sound waves, which travel by alternate compression and expansion of air (or of whatever substance is transmitting the sound), light must travel in transverse waves. That is, the waves must be at right angles to the direction of travel. But there were problems. If light was wave motion, what was it that traveled?

Ether was the answer. But it was a catch-all answer. Theory said that in order for waves to travel transversely, as light was believed to, the either had to be elastic, it had to be everywhere, and it had to have many of the properties possessed by solids. Could there be such a substance? Finding an answer posed many challenges.

Gradually, solutions were found. The presence of magnetic and electric fields in space, which were the concern of Michael Faraday (1791–1867) toward the middle years of the nineteenth century, revealed part of the answer. This partial answer led to the work of James Clerk Maxwell (1831–1879). In the latter half of the century he developed the electromagnetic theory of light. This put science well on its way toward finding answers to the challenging questions that had been recurring.

4 Light and Electromagnetism

Light is invisible. This seems contrary to what we observe, but suppose light were visible—suppose it were some sort of milky substance. We would find it impossible to see a tree, for example, because of the "milky light" that filled the space between us and the tree. It is only because light is invisible that we can identify light sources. We can control this invisible light with mirrors and lenses. Long before there were such controls, people were aware of the presence or absence of light. They could produce shadows, could see the rainbow of colors. Light gave them an awareness of the world, even though light itself was not understood.

Light was not the only mystery in the world of the ancient Greeks and Phoenicians. Lodestones, or "leading stones," as they were called, also caused wonderment. There was a "force" in a lodestone. If a piece of paper were placed on one, and iron filings sprinkled on the paper, the "force" acted through the paper. The behavior of a needle also proved the presence of the force. If a needle were placed on a lodestone, the force went right through the needle. Some of it remained, for if the needle were hung by a thread, the needle took a north-south line.

Long before there were steel needles, lodestones had served to guide mariners. The force within a lodestone could act through space. For centuries lodestones floating on bits of cork served as compasses for mariners sailing trade routes between ancient civilizations.

The force, whatever it might be, was there in the lodestone. Men were curious about it. But they were at a loss to explain it. They could not understand how it could flow from the lodestone to the needle, or why a free-moving lodestone took a north-south line.

In England during the late 1500s, Sir William Gilbert (1544-1603) studied the force, and made discoveries about it. When a needle was pivoted at the middle so it could swing vertically, the needle did not point straight up and down. In the northern hemisphere the one end of the needle pointed toward the North Pole of the earth–the North Magnetic Pole, as we now know it to be. Experiments with the "dipping needle," as the vertically swinging needle came to be called, and with compass needles, convinced Gilbert that the earth itself was a lodestone. It was a magnet—a Greek word for the stones found in Magnesia, which was a region of Asia Minor.

While studying lodestones, Gilbert mistakenly believed that gravity was another sign of the force of a magnet. We now believe that magnetism is produced as solid layers in the earth's interior. Electricity is generated, and the electricity is related to magnetism. Magnetism is believed to be electrodynamic, and so in no way like gravity, which is related to mass.

Gilbert's investigations aroused the curiosity of other investigators, one of whom was Johannes Kepler. Kepler said that while magnetism and gravity are not alike, gravity is another example of "action at a distance." Objects affected by gravity need not be touching one another—just as objects affected by magnets need not be touching. This idea of Kepler's was to influence Isaac Newton's investigations which led to his discovery of the laws of gravitation.

A basic idea about magnets which was discovered by William Gilbert is that there are different charges, or poles. Also, the two charges always exist together; one cannot be separated from the other. Suppose you were to file a nail almost in half and then stroke the nail a few times with a magnet. The nail will become a magnet. If the nail is then broken into two parts, both parts will be magnets and each magnet will have both a north and a south pole.

Gilbert also discovered that there is a "field" around a magnet. This field is a region, strongest near the poles, in which there is magnetic force. The region can be revealed by sprinkling iron filings on a piece of paper placed over a magnet. Each filing becomes a magnet that affects other filings, causing all to line up along the lines of force that compose the magnetic field. By using two magnets, with north and south poles, together but separated slightly, the attraction of the two can be seen. By reversing one magnet so that north and south are close together, repulsion of the fields can be shown.

Sir William Gilbert's magnets and their behavior were interesting. But they remained little more than curiosities for more than a hundred years. In the decades after his death in 1603, astronomers, biologists, physicists, scientists of many disciplines, moved in other directions. Galileo saw a new world through his telescopes, Kepler explained motions of the planets, and many investigators concerned themselves with the curious behavior of amber, the fossilized form of pine resin. When rubbed with silk, amber had the unique ability to attract light bits of dried plants. This ability, another example of "force at a distance," came to be called electricity, after *elektron*, the Greek word for amber.

The force could be "bottled up," it was found. A rod of amber was rubbed, then touched to a metal post inserted through the cork of a bottle. When this was done over and over again, the force was held in the bottle. When the force was discharged by bringing another metal rod close to the one in the bottle, a spark was created. The display was spectacular and people clamored to see it. However, electrostatics (as such phenomena came to be called) have little more to add to our story than to give us the origin of the word "electricity."

Of considerably more importance was the work of Luigi Galvani (1737–1798) and his fellow Italian Alessandro Volta (1745–1827). While experimenting with frogs, Galvani discovered the principle of the electric battery. When a nerve in a frog's leg was touched by an electric discharge from one of the jars mentioned above, the leg twitched. Also, when two wires of unlike metals—of zinc and copper, for example—were joined together and the free ends of the wires touched to the nerve, the leg twitched. Galvani thought he had discovered "animal electricity," as he called it. Rather he had found the idea behind a battery: two different metals in a liquid that conducts electricity (the body fluids of the frog's leg) will produce an electric current. In modern batteries (dry cells) the metals are often carbon and zinc and the "liquid" is a paste containing manganese dioxide.

Galvani's idea was developed by Volta when he connected several cups of brine with alternating strips of zinc and copper. Before this, electricity had been available only in one instantaneous discharge from a jar. Volta's battery of cells produced a flow of electricity. It opened the way to an equally tremendous discovery: the connection between magnetism and electricity. Whenever there is a movement of electricity, magnetism is produced.

The discovery came two hundred years after William Gilbert's work with magnets. In 1819, Hans Christian Oersted (1777–1851), a Danish teacher and scientist, found that a compass needle (a magnet) turns when a wire carrying electricity is held above it. The needle is not attracted to the wire, nor is it repelled. The needle is turned to the right or left—it is deflected. When the direction of flow of the electricity is changed, the needle changes direction.

This discovery was another amazing curiosity, of great interest to laymen and scientists alike. André Marie Ampère (1775–1836) experimented with the idea. So did another Frenchman, Dominique François Jean Arago (1786–1853). Ampère experimented with the "field" around a wire. He found that it surrounded the wire and, further, that the field can be made greater by coiling the wire. This had the effect of adding wires. When a piece of iron was put inside the coil, the field was concentrated. An electromagnet was the result. When current flows through the wire, the coil and the piece of iron become magnets. Arago continued to study the field around the wire. His work inspired the English scientist Michael Faraday to investigate the work that had been done by Ampère. Faraday performed many experiments that resulted in the momentous discovery relating electricity, magnetism, and light. This discovery was explained mathematically

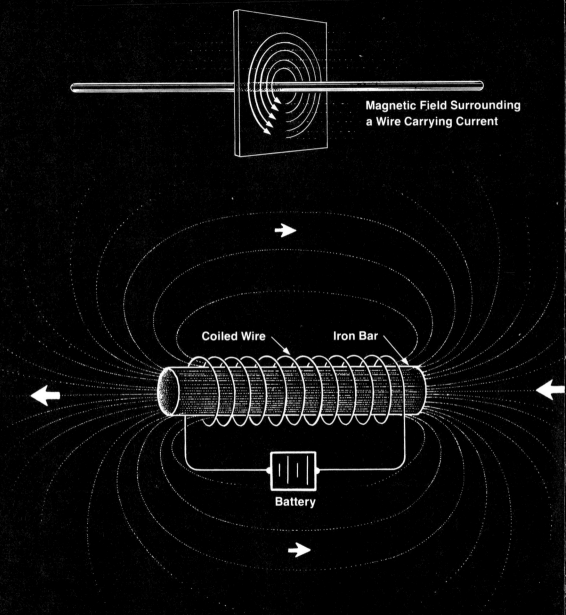

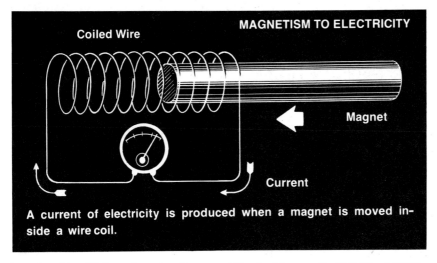

A current of electricity is produced when a magnet is moved inside a wire coil.

by the English scientist James Clerk Maxwell (1831–1879).

Faraday found that when electricity in a coil is turned on/off/on, an electric current also flows in a second coil. The second coil may be near to the first, or it may surround the first. Also, he found that if a magnet is moved into and out of a coil of wire, an electric current flows in the wire. The idea of the electric generator was born, and also the principle of the electric motor. If there was any single discovery that was to usher in the "age of electricity," this was it. Here was the way to change one kind of energy of motion (falling water, a turning wheel—as in a steam engine or turbine) into another form of energy.

Faraday was fascinated by the fields with which he had been working. He found that fields exist in the space between coils of wire, or in the magnetic field around a single wire. Also, Faraday reasoned, light must exist in space, and must be similar—or at least one must affect the other. His theories led him to another important discovery.

Faraday placed a piece of glass between the poles of an electromagnet. He then passed polarized light, light that had gone through tourmaline, through the glass. The light then went to a second polarizer that was turned so as to extinguish the light. When the electric current was turned on, the second polarizer had to be rotated to a new position to turn off all the light. Obviously the magnetism produced by the current was affecting the light as it passed through the glass plate. Faraday concluded that light is similar to an electromagnetic field (or "stress" as Faraday called it). Also, Faraday was convinced that these stresses could be transmitted as waves through the ether.

The idea was defined later by Maxwell. In a paper published in 1865 he wrote: "we have strong reason to conclude that light itself (including radiant heat, and other radiations if any) is an electromagnetic disturbance in the form of waves. . . ." In this same paper he said that if electromagnetic "stresses" were produced they would travel at the speed of light. Also, they would travel in transverse waves, the waves being produced by periodic vibrations of electric and magnetic stresses.

A generation later the stresses discussed by Faraday and Maxwell were produced in a laboratory. It was established that we are surrounded by electromagnetic phenomena—light being only the one the most apparent to our senses.

5 Electromagnetic Waves

By the middle of the nineteenth century a great deal was known about light, magnetism, and electricity. Centuries earlier Thomas Young had measured the wavelength of light, William Gilbert had discovered the polarity of magnets, and numerous investigators had experimented with the new wonder—electricity. In 1865, Maxwell theorized about electromagnetic waves, but he was unable to provide specific directions for producing the waves. He did say that if an oscillating current (one that changes direction rapidly, flowing first one way, then the other) were set up, an electromagnetic wave would result.

Long before this time the frequency of light had been determined. It would seem, then, that there was a way to prove Maxwell's theory. Set up an oscillating current—one with a frequency of 4×10^{14} cm/sec (the frequency of red light)—and see if red light was given off by the oscillator.

The reasoning was correct. But no one knew how to set up a current that oscillated at that frequency. In fact, no one has ever found a way to do so. Even with the most modern equipment, frequencies cannot be pushed much beyond 10^{12} cm/sec. This is much below the frequency of visible light—4×10^{14} cm/sec, for red light, to 8×10^{14} cm/sec, for violet light.

There seemed no way to observe the electromagnetic waves that might be produced by an oscillator as Maxwell had suggested. Therefore interest in them lagged for the next twenty years or so. However, in 1887, Heinrich Hertz (1857–1894), a German physicist, found a way to prove their existence.

Hertz set up two metal plates and fastened a metal rod to each plate. On the end of each rod was a polished brass sphere. The two spheres were separated from each other by a few millimeters. Hertz charged the plates with electricity. When they could hold no more, the plates discharged. Electricity flowed back and forth between the plates some ten to a hundred million times in one second. When that happened, an electric spark was produced.

As the current flowed back and forth (oscillated) between the two plates, Hertz suspected that an electromagnetic wave had been generated. To find out, he invented a detector. It was a hoop of wire about a meter across. The hoop was cut open at one point and small polished brass knobs fastened to the ends of the wire. The distance between the knobs could be adjusted to a millimeter or so. To use the detector, Hertz placed the loop horizontally, with the knobs of the oscillator and those of the dectector a meter or so apart. When the detector was adjusted properly and the oscillator was activated, sparks jumped between the knobs of the detector. Yet there was no connection between the detector and the oscillator.

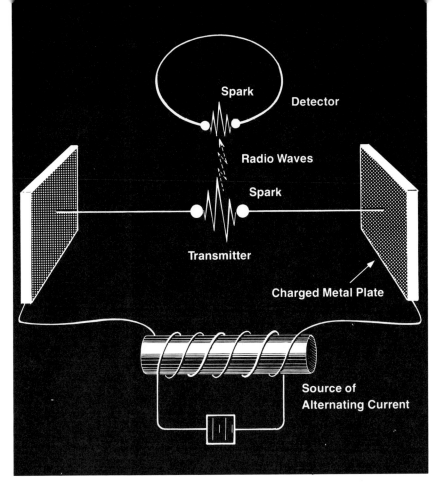

Radiant energy in the form of radio waves is produced by the spark made by the transmitter. The waves travel through space and charge the detector. When the detector discharges, another spark is produced.

The experiment was a milestone in the growth of understanding of the electromagnetic spectrum, for it proved that waves could be generated and that they traveled through space. They could not be seen because the frequency was far below that of visible light. The waves were often called Hertzian waves, and still are by radio engineers. They were radio waves, as we know them today—waves that radiate through space.

Hertz expanded his experiments and discovered that the waves pass through glass, wood, or even a brick wall. He built a prism out of asphalt and found that the waves were bent as they passed through, much as light waves are bent by a glass prism. Furthermore Hertz measured the length of the waves that he produced. He found them to be 9.6 meters, indicating that they had a frequency of 31 000 000 oscillations per second. (You can understand why the waves could not be seen–red light has a frequency of some 400 000 000 000 000 oscillations per second.)

However, there were enough similarities between these Hertzian waves and light waves to conclude that the two were related. They differed only in frequency and wavelength, the Hertzian waves being much longer. Hertz himself said: "The described experiments appear, at least to me, in a high degree suited to remove doubt in the identity of light, heat radiation, and electrodynamic wave motion."

This being so, it would seem that light, too, must be caused by some sort of oscillation. Perhaps particles contained in the light producer oscillated and so produced visible light.

In 1896, two years after the death of Heinrich Hertz, scientists found that the theory was correct. When a flame was placed between the poles of a strong electromagnet, the color of the light changed slightly. The magnetic waves affect the electric charges in the particles producing the light. This causes their frequency to change.

Hertzian waves are produced by relatively slow oscillations of an electric field. The oscillations generate long waves— several meters in length. Light waves may be produced in the same way, by rapid oscillations of an electric field in the particles of the light producer. The oscillations, which are high frequency, generate short waves of only a few ten-millionths of a meter.

After centuries of wondering and experimenting, scientists had uncovered the nature of light. It is an electromagnetic phenomenon. Apparently oscillations of atoms (or parts of atoms) produce it.

With this knowledge, new and even more puzzling questions appeared. What is an atom composed of? What happens inside an atom? How does an atom produce radiation? What keeps it going? Finding answers demanded the utmost efforts of such scientists as Albert Einstein, Niels Bohr, Ernest Rutherford, Max Planck, and scores of other investigators.

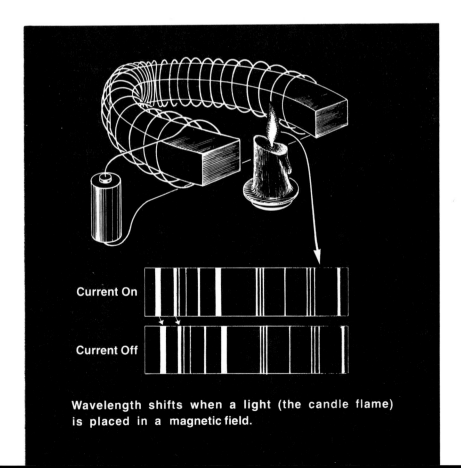

Wavelength shifts when a light (the candle flame) is placed in a magnetic field.

6 Max Planck: Packets of Energy

In the early 1800s it was known that there was more to the spectrum of the Sun than Newton's band of visible colors. Some time earlier Sir William Herschel, who is best known for his sky survey and the discovery of Uranus, had placed a thermometer beyond the red end of the visible spectrum. Temperature increased, indicating that heat was present. Other investigators had proved that there was energy beyond the violet end of the spectrum. The rays there darkened silver chloride salts more strongly than did the violet light. These rays, called ultraviolet because they lie beyond the violet region of the spectrum, were originally called the "dark chemical rays."

Experimentation with the spectrum of the Sun revealed that the wavelength of visible light lay between about 4 000 angstroms (1A = 10^{-8} cm), presently expressed as 400 nanometers, and 7 500 angstroms, or 750 nanometers). (The angstrom, a unit of length, is named after Anders Jonas Ångström (1814–1874), a Swedish physicist.) At either end there was additional radiation, but it could not be detected visually. Ultraviolet light was beyond the violet, and infrared ("infra-" meaning "beneath" or "below") below the red.

Suppose an iron nail is heated electrically. Before any color can be seen, the nail gives off infrared radiation. You know this because the radiation can be felt as heat. If you were to photograph the nail, using infrared film, a sharp image would be produced. Additional heating causes the nail to give off shorter wavelengths. These can be seen as red light; the nail has become red-hot. Continued heating causes the color of the nail to change to orange, yellow, and so on, until the nail becomes white-hot; it is giving off a variety of wavelengths. If the white light were passed through a diffraction grating, the various wavelengths would produce a continuous band of colors. And if suitable detectors were used, the band would be found to continue into the infrared and ultraviolet zones. The ultraviolet radiation can be measured by the effect it produces on a photoelectric cell (a device made of selenium or a similar metal that produces electricity when light or other forms of radiation fall upon it). When the cell is placed beyond the violet, electricity continues to be generated. The infrared radiation may be detected with a thermocouple. This is made of two different metals twisted together; electricity is produced when radiation falls upon them.

If the nail could be heated to 6 000°K—the temperature of the solar surface—the distribution of wavelengths would be as shown in the illustration. The peak of energy is in the green wavelength. However, the greatest amount of energy is in the infrared region. Our eyes combine the many different wavelenghts, receiving them as white light.

As the temperature of a solid, a liquid, or a dense gas increases, more energy is associated with each wavelength. Also, as temperature increases, the wavelengths are shifted toward the ultraviolet—toward the shorter wavelengths. This is true whatever the substance that is giving off the radiation. However, the rate of radiation varies from one material to another, depending upon the surface. A shiny surface will not absorb radiation, or give it off, nearly as rapidly as will a surface that is dull black. Also, a rough surface gives off radiation more rapidly than does a smooth one.

To study radiation and establish rules about it, scientists required a perfect emitter, one that would give off all wavelengths equally. They needed something that would absorb all, or nearly all, wavelengths of any radiation (heat, light, infrared, ultraviolet) that fell upon it. Also, once heated, the surface would give off all the energy it contained, and equally at all wavelengths.

No such surface has ever been produced. However, toward the end of the last century an almost perfect black body (as such a radiator is called) was made. It was a hollow black metal ball with a single small hole drilled into it. The ball, blackened inside and out, was then placed inside an oven.

When receiving radiation, the waves enter the ball through the hole. The radiation is scattered inside the ball and absorbed by the metal. A little escapes through the hole, but the total is very small. This is called black-body radiation.

The radiation is analyzed at different wavelengths and the information is plotted on a chart. The curve that results is like the one shown on page 54. With increasing temperature, the curve moves to the left (toward the ultraviolet), as expected.

This almost perfect absorber-emitter of radiation made it possible for scientists to test laws that had been arrived at theoretically. One was that the amount of energy given off is related to the fourth power of the temperature. This means that a slight increase in temperature results in a considerable increase in the energy given off.

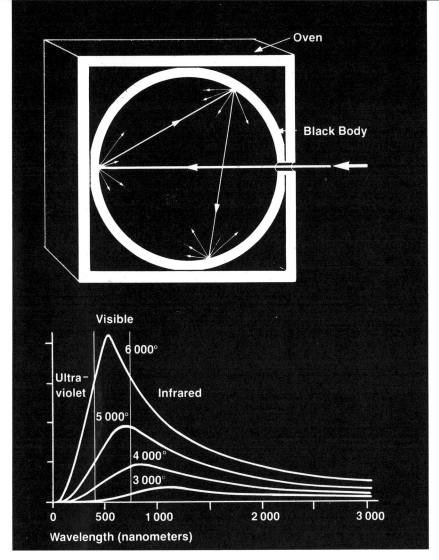

When radiation enters the ball, the black body acts as a perfect absorber. When heated, the black body gives off radiation at all frequencies; it then acts as a perfect emitter.

Another law that was verified relates the peak wavelength to a specific temperature. This means that the temperature of the Sun, for example, could be determined by analyzing the solar spectrum. The highest peak is reached in the green area, with a wavelength of 500 nanometers. When this figure

52 THE ELECTROMAGNETIC SPECTRUM

is used in an equation, the temperature of the Sun is found to be 5 800°K.

This is a bit below the 6 000°K determined for the Sun in other ways. The difference is because some of the radiation is absorbed by earth's atmosphere. Also, the Sun is not a perfect black-body radiator.

The peak wavelength of any star, nebula, or an entire distant galaxy could be used. Applying it in the same equation, the temperatures of those stars, nebulas or galaxies could be determined.

QUANTA OF ENERGY

Laboratory experiments with black-body radiators revealed considerable information. But scientists were unable to explain what was happening; why did the radiated energy always produce the curve that was observed—why did wavelengths shift toward the violet as energy increased? It was not known why or how the energy was given off. That is, not until Max Planck (1858–1947), a German scientist, introduced some radical ideas that provided new directions in thinking and experimenting.

At the close of the year 1900, Planck startled the world of science. He said that a body gives off energy in packets, or quanta (a single packet would be a quantum). The body would lose energy (cool) unless it was resupplied—that's what the oven did for the experimental black body. The packets were always whole units. There could be 2, 3, 4, let us say, but not 2½, 3¼, and so on. Energy was lost in steps, not smoothly. The process could be better represented as a descending staircase rather than as a smooth ramp.

If this were so, it was argued, how could you explain a pendulum? A weight swinging on the end of a string contains energy. It is an oscillating body, one that swings back and forth. As a pendulum swings, it slows down. But it does not slow down in small jumps—from one stair, for example, to the next. It does not appear to lose energy in quanta—whole units that drop it abruptly from one level to another. Rather the pendulum slows down smoothly and evenly. The question was a dilemma to Planck as well as to other scientists at the turn of the century. The answer, it was found, has to do with the amount of energy involved.

By using mathematics rather than observations, Planck found that the energy given off by a radiator (a black body) was determined by the frequency multiplied by what has become known as Planck's constant. It is 6.625×10^{-34} joules per second. (A joule is a very small amount of energy. Someone has said it's the amount of energy contained in a mosquito flying full speed ahead. And Planck's constant is .000 000 000 000 000 000 000 000 000 000 6625 of that —an infinitesimally small amount of energy.) It is so small that single changes cannot be observed even though they are abrupt, because the change is always made in whole units. So, even though the pendulum does slow down in jumps, there are so many and the change from one to the next is so slight one cannot observe them. The pendulum appears to slow down evenly. In effect, Planck's constant has meaning only for submicroscopic systems in which energy changes are extremely small.

Before Planck's time, Maxwell had established that energy radiates in waves—that energy is a wave phenomenon. And Planck accepted that idea. But, said Planck, the process is more complicated than Maxwell had made it out to be—the waves are made of quanta of energy. Submicroscopic systems, (groups of atoms, for example) give off radiation at all

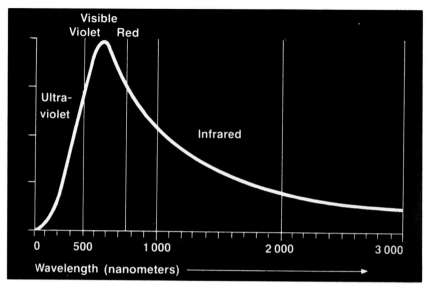

Groups of atoms (the Sun, for example) give off radiation at all frequencies. Energy distribution is represented below the curve. In this case the greatest energy is in the visible area.

frequencies. When the frequencies are charted, a Planck curve is produced. The region below the curve represents the total amount of energy. The high point shows the peak frequency. A Planck curve is shown above.

Planck's theory maintained that the oscillator (a vibrating body) does not give off energy continually. It does so only when the amplitude (which would be loudness if we were considering sound waves) of the vibrations changes. This idea disagreed with the theories of Maxwell. And there were other problems.

It was not clear how radiation from an oscillator could travel through space in waves, and be picked up by another oscillator. How could the receiver gather in the quanta of energy needed for it to change to the next higher energy level? Did it add up energy until the necessary amount was reached? Why did the energy given off always produce a Planck curve of distribution? Also, it was known that radiation from solids was different from the radiation given off by liquids and gases—or it seemed to be different. Planck's quantum theory could not explain why this should be so.

At the start of this century scientists felt that the theory was interesting but not especially important; there appeared to be too many weaknesses. However, in the next few years Planck's ideas were proved correct by the work of Albert Einstein, who was experimenting with the relationships between light and electricity, and Ernest Rutherford and Niels Bohr, who were studying the atom and its structure and motions.

7 Albert Einstein: The Photon Theory

Some ten years before Max Planck startled the scientific world with his quantum idea, Heinrich Hertz had shown that Maxwell was right when he said that light is an electromagnetic radiation—one that travels in waves. You recall that Hertz transmitted a wave generated by a spark and picked up the wave with a receiver. At the same time Hertz set the stage for Planck's quantum idea. He determined that ultraviolet light from the spark caused electrons to be given off by the metal sphere. The electrons helped to maintain the current between the spheres. The air around the spheres did not have any effect on the action. This was proved by directing ultraviolet light onto a metal surface, comparable to one of the spheres, which was inside a glass tube from which the air had been removed. The metal gave off electrons which passed through the vacuum and were picked up by another metal plate—comparable to the second metal sphere. The process of giving off electrons when stimulated by light is called the photoelectric effect.

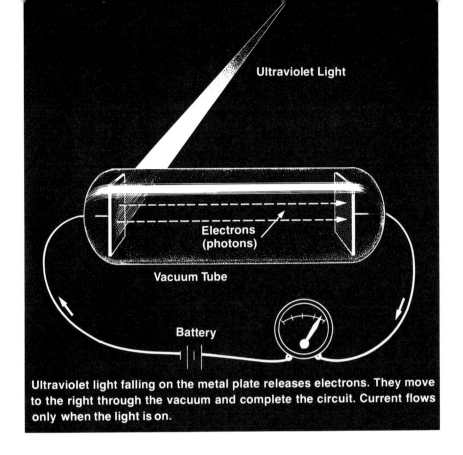

Ultraviolet light falling on the metal plate releases electrons. They move to the right through the vacuum and complete the circuit. Current flows only when the light is on.

When the light is made more intense, the number of electrons that are given off becomes greater. But the energy level of each electron does not change. The only way the electrons can be made to carry more, or less, energy is by changing the frequency of the light. For example, blue light releases more energetic electrons than does red light.

According to Maxwell's wave theory of light, energy is spread evenly over the entire wave. It was known that a certain amount of energy is needed to release an electron from a plate or sphere. Also, it was supposed that the electron gathered in energy. The energy was added up and the electron set free only after sufficient energy had been accumulated. If this were so, there would have to be a delay between the time the light was turned on and the electron was set free. But the action was instantaneous. Apparently the theory of accumulation of energy was incorrect. Some other explanation was needed.

In 1905, Albert Einstein (1879–1955), a German scientist who later became an American citizen, suggested that light and other forms of radiation should be thought of in two ways. Maxwell's wave theory, he said, made it entirely possible to explain reflection, interference, polarization, and other such phenomena. But to explain the photoelectric effect, the energy in a light wave must be distributed unevenly. The energy is not spread evenly along the wave but in bundles, or packets. Einstein's packets were later to be called photons.

Photons have no mass and so it is difficult to comprehend their existence. They are bundles of energy, said Einstein, that stud each wave that is given off by a glowing object—the Sun, for example. The beams of light that are used in many light experiments (those involving reflection, interference, and such phenomena) contain so many photons that the energy appears smoothed out and evenly distributed. (In much the same manner objects appear to us as continuous matter even though we know that they are made of separate atoms and molecules.) However, when submicroscopic particles (electrons) are involved, the effect of individual particles of the light wave (the photons) can be noted.

The theory enabled scientists to answer some of the questions about the photoelectric effect. There was not time delay because electrons did not accumulate energy from a light wave. Rather the electrons absorbed only those photons that had a certain amount of energy—the amount needed to set the electrons free. Also, the energy level of the electrons cannot be increased by increasing intensity, but only by increasing frequency. When frequency of the light is increased, the photons are more energetic; ultraviolet photons, for example, are much more energetic than are infrared photons. When these more energetic photons strike the plate, they are absorbed by electrons more deeply imbedded in the plate. This greater energy is transferred to the electrons.

You recall that Max Planck's quantum idea was not widely accepted when first presented. However, that attitude began to change after Einstein proposed the "photon" explanation for the photoelectric effect.

Still there were dilemmas.

How could light be two things at the same time; how could it be a wave and also be made of particles? The problem is not as great as it seems, providing you can think of a photon as a concentration of energy and not as a pelletlike bundle of matter. Also, as mentioned earlier, you might think of a light wave (or other energy wave) as being made of so many photons that they blend together. They might be compared to the drops of water which all together make an ocean wave that appears to be continuous. Under certain conditions light reveals its wave structure, and at other times it produces reactions that can best be explained by the photon structure.

Perhaps you'll have a better idea of the behavior of photons if we consider some of the ways they have been harnessed.

FLUORESCENT LIGHT

A fluorescent tube often contains mercury, at low pressure. When the current is turned on, electrons bombard the mercury atoms. Electrons in those atoms take up the energy and immediately give it off again as ultraviolet, invisible radiation. The inside of the tube is coated with a fluorescent or phosphorescent metallic salt (usually a calcium salt or one of zinc, such as zinc sulphide or zinc silicate). The ultraviolet radiation falls on the coating. This causes the molecules to eject energy in the form of visible light, at a lower energy level than the ultraviolet light. The energy level (color) of the light given off by the lamp is determined by the kind of coating on the inside wall. Each energy change, from electron to ultraviolet to visible light, is achieved in sharp, stairlike jumps. However, because of the large number of atoms involved, the changes appear to be continuous.

X-RAY (ROENTGEN) TUBES

Wilhelm Konrad Roentgen (1845–1923) was a German scientist who conducted many experiments with electricity. In 1895 he was working with electricity and vacuum tubes. When he turned on the current, he noticed that a nearby screen coated with barium salts began to glow. Fascinated with what he had seen, Roentgen investigated further and found that, unlike light, whatever was causing the glow could go through metal. He was unable to understand this curious radiation and so called it X radiation, or X rays. The name remains—although "Roentgen rays" is also used.

In its simplest form, an X-ray tube is a vacuum tube with a cathode and an anode. (A cathode releases electrons when it is heated, while an anode attracts electrons.) When turned on, the cathode of the tube releases high-energy electrons. Some of the energy in these electrons serves to heat up the anode. Other electrons that strike the anode cause it to give off X-ray photons. These pass through the glass readily. And, as we know, they can pass through many other substances.

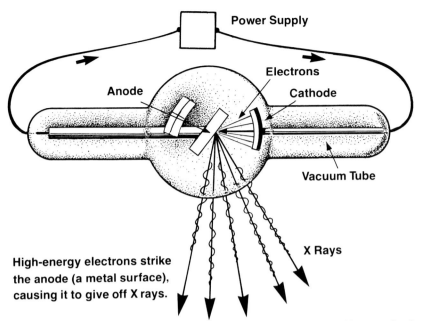

High-energy electrons strike the anode (a metal surface), causing it to give off X rays.

When Einstein experimented with photons he directed ultraviolet light onto a target. This resulted in the production of photoelectrons and an electric current. In a Roentgen tube the action is reversed. Here an electric current is used to produce radiation in the form of very energetic photons—energetic enough to be classified as X rays.

As energy increases, wavelength becomes shorter. Low-energy photons are associated with long-wavelength radiation, as tabulated below:

RADIATION	WAVELENGTH RANGE
Radio waves	A few meters and up
Microwaves	A few millimeters to meters
Infrared radiation	750 to .01 mm
Visible light	380 to 750 nm
Ultraviolet light	10 to 380 nm
X rays	.01 to 50 nm
Gamma rays	Less than 0.05 nm

COMETS AND PHOTONS

The tail of a comet always points away from the Sun. This is easy to understand when a comet is approaching the Sun. The problem becomes complicated when explaining why the "tail" should go ahead of the comet, when the comet is moving away from the Sun, much like the beam of a flashlight.

In the early seventeenth century Johannes Kepler had suggested that sunlight pushed away the comet's tail. At that time the idea seemed ridiculous to a good many people. However, careful measurements made two-hundred years later indicated that Kepler could be right. It was found that light does exert pressure; light does have momentum.

The equation for finding the momentum of anything involves the mass of the object and its velocity. If light has momentum, it must have mass. And if it has mass, then it must be made of particles, or quanta, as Planck said. The action of sunlight on the gases of a comet justified Einstein's idea of photons. In addition, we now know that subatomic particles ejected by the Sun (the solar wind) affect the nebulous gases that compose a comet.

Using Einstein's photons and Planck's quanta, it became possible to explain the radiation of energy from the Sun and by other collections of gases. Also, it became possible to describe in more detail the way in which an atom is put together.

8 Atoms and the Spectra of Gases

As early as 1752 it was known that a gas produces a spectrum quite different from that of a solid. In that year the Scottish scientist Thomas Melvill placed various substances in a flame. When he put a prism between himself and the light and then observed through a small round hole, Melvill saw a series of round spots of color. He recorded that "there were individual circular spots, each having the color of that part of the spectrum in which it was located and with dark gaps—missing colors—between the spots." Today a slit would be used for viewing rather than a round hole, and so lines are observed rather than round spots. The lines produce a bright-line spectrum, which we touched upon in Chapter 2. It is often called an emission spectrum, or emission line spectrum.

Melvill found that the array of colored dots was always the same for any given substance. Sodium, you recall, produces two lines in the yellow region; neon produces scores of lines, mainly in the red area. Most of the scores of lines produced by nitrogen are in the blue region. The lines produced by each substance are so definite they can be used to identify the substance. The emission line spectrum of a material is unique, just as your signature is unique, or your fingerprints. In some cases the spectrum is very simple, as in the case of sodium. More often the spectrum of a substance is complex. When iron is vaporized it produces over 6 000 bright lines —always the same lines. But why? Investigators wondered what there was about substances that produced the individual lines. How did substances differ to such an extent that the spectra they produced were so different from one another? In seeking answers, scientists considered and studied the achievements of one another.

DARK-LINE SPECTRA

In the early part of the nineteenth century, Joseph Fraunhofer (1787–1826), the German scientist who had invented the diffraction grating, was observing the solar spectrum. He noticed there were several hundred dark lines arrayed across the continuous range of spectrum colors. When he studied the light from other stars many, though not all, of the same lines appeared. These dark spectral lines have since been called Fraunhofer lines.

Later on Gustav Robert Kirchhoff (1824–1887) produced some of these dark lines in a laboratory. He passed light from a hot solid (which makes a continuous spectrum) thorough a gas at lower temperature and then through a prism. A continuous spectrum with dark lines upon it was produced. The dark lines were at the same locations where bright lines appeared when a spectrum of the gas alone was produced. Kirchhoff concluded that the cooler gas absorbed certain wavelengths.

Since the Sun produces dark lines, said Kirchhoff, the Sun must be surrounded by an atmosphere—by gases that are cooler than the interior and which are of low density. What's more, by analyzing the lines we should be able to determine what gases are in the solar atmosphere. All of this reasoning was correct. Today some 67 elements have been identified in the solar atmosphere by spectroscopic analysis of sunlight.

At the time of Kirchhoff's death there was wide activity in spectroscopy. Scientists were discovering the bright lines emitted by multitudes of materials. But several basic questions remained unanswered: Why did gases emit radiation and absorb it only in certain ways? Why were spectral lines spaced as they were? There seemed no pattern to the arrangement—or was there?

There had to be explanations. In the 1880s scientists tried to find a mathematical way of predicting where the lines of a substance would appear. A Swiss schoolteacher named Johann Jakob Balmer (1825–1898) found the answer.

The positions of four of the many lines of hydrogen had been measured accurately some years earlier by Angström. Balmer discovered a formula that enabled him to determine those positions. By using his formula Balmer could predict the wavelength of each of the lines. The table below shows the wavelengths measured by Angström and also the wavelengths predicted by Balmer's formula. You can see how closely they agree.

| | WAVELENGTH | |
LINE	ANGSTRÖM MEASUREMENT	BALMER PREDICTION
H alpha (red)	6562.10	6562.08
H beta (green)	4860.74	4860.08
H gamma (blue)	4340.1	4340
H delta (violet)	4101.2	4101.3

Balmer had opened the way toward mathematical explanation of spectral lines. Others were to carry the work further. Outstanding among them were the German scientist F. Paschen, who predicted lines located in the infrared region, and Theodore Lyman, who made predictions of lines that were later found in the ultraviolet area.

Although the formulas that were developed made it possible to predict the locations of lines in the spectra of substances, scientists still did not know why the formulas worked. What was happening in the substances to cause them to produce emission lines? Answers could not be found until the early 1900s. One needed first to understand the quantum ideas of Max Planck and the photoelectric effect that had been explained by Einstein. Also needed was information about how atoms are put together.

EARLY THEORIES ABOUT ATOMS

After it was known that electrons existed, a theory to explain how an atom gave off light was proposed. The atom was considered to be a globule of charged liquid. Electrons were imbedded in the fluid, somewhat like raisins in a cake. When the atom was excited, it was reasoned, the electrons were pushed aside. The charged globule would pull them back. In this way the electrons were made to oscillate. It had been shown that an oscillator would produce radiation. In this case the radiation would be in the form of visible light.

The theory was respected. However, it had flaws. For example, it could not explain why the light was discrete—why the spectrum was made of separate bright, sharp lines. If it resulted from oscillation, the light should be continous; the spectrum should be a band of colors, each one blending into the next.

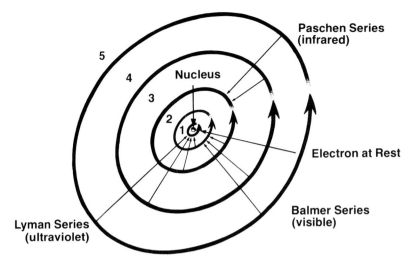

Five possible orbits of the hydrogen electron. Spectrum lines are produced when the electron shifts orbits. They may be in the infrared region when there is low energy (Paschen series), in the visible light area (Balmer series), or ultraviolet when there is high energy (Lyman series).

A host of investigators made giant strides toward finding models of the atom that would make possible the production of the observed line spectra. Among these scientists were Ernest Rutherford (1871–1937), an Englishman, and Niels Bohr (1885–1962), an Dane. They established that the atom was not a globule like a melon with seeds (electrons) in it. Rather it was more like a miniature solar system: a small massive nucleus with electrons in orbit around the nucleus. The electrons moved in definite orbits, which represented different energy levels. When the hydrogen atom, which has only one electron, was at rest, the electron moved in orbit number one, let us say. Should the hydrogen atom be placed in a glass tube and an electric current passed through the tube, the electron would pick up energy. The electron would move from orbit number one to two, or three, or higher—depending upon the amount of energy that the electron absorbed. It could not go to energy level (orbit) 1½, 2¼, or to any fractional level. It must always move in whole steps.

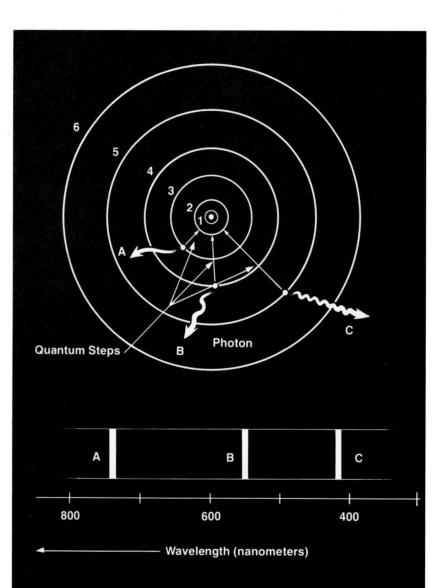

When an electron moves to a lower-energy orbit, it releases quantum steps of energy (photons) having various wavelengths — from A (the longest) to C (the shortest).

When the electron picks up energy, the atom is said to be charged, or excited. Immediately the electron will fall back to a lower orbit, seeking the unexcited state. As the electron falls back, it gives off energy. But it does not emit energy steadily—it does so in bursts. The amount of energy is a quantum step (a Planck quantum); the energy itself is a photon, as explained by Einstein. Dozens of energy levels are possible in the hydrogen atom, the simplest of all the atoms. In the more complex atoms—iron, for example, which has 26 electrons—there are thousands of possible energy levels. The photons emitted from the excited atoms have specific wavelengths, which means they produce specific bright lines. Some of these will be in the range of visible light; some may be in other regions of the spectrum. The number of electrons in atoms ranges from 1 for hydrogen to 105 for hahnium (named after Otto Hahn, who studied the atomic nucleus). The various atoms (elements) are displayed in the periodic chart on page 70.

Knowledge of the atom made great strides in the 1930s. The nucleus was explored intensively. Its energy was set free in the nuclear bomb, first exploded in 1945, and is now being liberated in electric generating stations powered by nuclear fission. Hans Bethe (1906–) applied the new-found knowledge of the atomic nucleus to the Sun when he explained how energy was generated there, and in other stars.

Scientists who lived at the close of the nineteenth century and the early years of the twentieth century—Angström, Fraunhofer, Balmer, Paschen, Rutherford, Planck, Einstein, to name only a few—made it possible for us to understand the atom and the electromagnetic spectrum, the key to understanding the universe.

The Chemical Make-up of the Universe

Period	Group IA	IIA	IIIB	IVB	VB	VIB	VIIB	VIII			IB	IIB	IIIA	IVA	VA	VIA	VIIA	VIII
1	1 Hydrogen **H** 1																	2 Helium **He** 4
2	3 Lithium **Li** 7	4 Beryllium **Be** 9											5 Boron **B** 11	6 Carbon **C** 12	7 Nitrogen **N** 14	8 Oxygen **O** 16	9 Fluorine **F** 19	10 Neon **Ne** 20
3	11 Sodium **Na** 23	12 Magnesium **Mg** 24											13 Aluminum **Al** 27	14 Silicon **Si** 28	15 Phosphorus **P** 31	16 Sulfur **S** 32	17 Chlorine **Cl** 35	18 Argon **Ar** 40
4	19 Potassium **K** 39	20 Calcium **Ca** 40	21 Scandium **Sc** 45	22 Titanium **Ti** 48	23 Vanadium **V** 51	24 Chromium **Cr** 52	25 Manganese **Mn** 55	26 Iron **Fe** 56	27 Cobalt **Co** 59	28 Nickel **Ni** 59	29 Copper **Cu** 64	30 Zinc **Zn** 65	31 Gallium **Ga** 70	32 Germanium **Ge** 73	33 Arsenic **As** 75	34 Selenium **Se** 79	35 Bromine **Br** 80	36 Krypton **Kr** 84
5	37 Rubidium **Rb** 85	38 Strontium **Sr** 88	39 Yttrium **Y** 89	40 Zirconium **Zr** 91	41 Niobium **Nb** 93	42 Molybdenum **Mo** 96	43 Technetium **Tc** 99	44 Ruthenium **Ru** 101	45 Rhodium **Rh** 103	46 Palladium **Pd** 106	47 Silver **Ag** 108	48 Cadmium **Cd** 112	49 Indium **In** 115	50 Tin **Sn** 119	51 Antimony **Sb** 122	52 Tellurium **Te** 128	53 Iodine **I** 127	54 Xenon **Xe** 131
6	55 Cesium **Cs** 133	56 Barium **Ba** 137	57 *Lanthanum **La** 139	72 Hafnium **Hf** 178	73 Tantalum **Ta** 181	74 Wolfram (Tungsten) **W** 184	75 Rhenium **Re** 186	76 Osmium **Os** 190	77 Iridium **Ir** 192	78 Platinum **Pt** 195	79 Gold **Au** 197	80 Mercury **Hg** 201	81 Thallium **Tl** 204	82 Lead **Pb** 207	83 Bismuth **Bi** 209	84 Polonium **Po** 210	85 Astatine **At** 210	86 Radon **Rn** 222
7	87 Francium **Fr** 223	88 Radium **Ra** 226	89 **Actinium **Ac** 227	104 Kurchatovium **Ku**	105 Hahnium **Ha**													

Atomic number → 11 Sodium **Na** 23 ← Name, Approximate atomic weight (to nearest whole number)

Lanthanide Series*

6	58 Cerium **Ce** 140	59 Praseodymium **Pr** 141	60 Neodymium **Nd** 144	61 Promethium **Pm** 147	62 Samarium **Sm** 150	63 Europium **Eu** 152	64 Gadolinium **Gd** 157	65 Terbium **Tb** 159	66 Dysprosium **Dy** 163	67 Holmium **Ho** 165	68 Erbium **Er** 167	69 Thulium **Tm** 169	70 Ytterbium **Yb** 173	71 Lutetium **Lu** 175

Actinide Series**

7	90 Thorium **Th** 232	91 Protactinium **Pa** 231	92 Uranium **U** 238	93 Neptunium **Np** 237	94 Plutonium **Pu** 242	95 Americium **Am** 243	96 Curium **Cm** 247	97 Berkelium **Bk** 247	98 Californium **Cf** 251	99 Einsteinium **Es** 254	100 Fermium **Fm** 253	101 Mendelevium **Md** 256	102 Nobelium **No** 254	103 Lawrencium **Lr** 257

Elements in the Sun

Elements in Cosmic Rays (and Sun)

Elements in Interstellar Matter (and Sun and Cosmic Rays)

Deuterium, an isotope of hydrogen, has also been detected in interstellar matter.

9 The Visible Spectrum

Visible light is made up of those wavelengths between about 400 nanometers (violet) and 700 nanometers (red). This is the range of radiation to which the human eye is sensitive. It is a very narrow band when compared with the entire electromagnetic spectrum. Yet through history until about a hundred years ago all our knowledge of the universe—great and small—was obtained in the form of visible light.

Until the seventeenth century it was obtained with man's eyes alone. Early in that century Antony van Leeuwenhoek (1632–1723), the Dutch scientist, took Hans Lippershey's lenses and arranged them to make a microscope. With it he could see the unseeable—the capillaries of the blood system, the structure of teeth and muscles, even bacteria ("little beasties," as he called them). About the same time Galileo used the lenses of the spectacle-makers to design telescopes, which he turned to the skies. Previous to that time information about the Sun, Moon, planets, and stars was derived from unaided-eye observing.

Galileo's telescope was a tremendous improvement, for it enabled the eye to gather in more light. The pupil opening of the eye is only about 3 millimeters across. All the light that we see must come through that narrow aperture. A telescope gathers in light from a wider area (the size of the objective, the light-gathering lens or mirror, of the telescope) and so makes it possible to see dimmer objects. The ocular, or eyepiece, of the telescope (the lens nearer the eye) enlarges the image produced by the objective. The result is a larger, brighter image than one could see without a telescope.

Galileo assembled many telescopes, or "optik tubes," as they were called in his day, and spent much time making improvements. His first telescope magnified 3 times; his best was 33-power. This is about equal to the power of the small, crude telescopes one often sees in toy stores. Slight flaws in the lenses, also variations in curvature, tended to obscure the images that were seen. Even so, Galileo was able to see the phases of Venus, four of Jupiter's satellites, the rings of Saturn. The latter he could see only poorly. In fact, the rings appeared to him as extensions on either side of the planet. Not knowing their true explanation, Galileo said, "Saturn has ears."

The "optik tube" caused great excitement in the 1600s. Scientists throughout Europe visited Galileo for a look through his telescope. When possible they bought a telescope from Galileo or went home and set about constructing their own. Larger telescopes were made. The lenses became more perfect in shape, and the glass less flawed. All of the early telescopes were refractors: light was controlled by passing it through a simple lens (or lenses) that refracted (bent) the light.

A serious drawback of Galileo's telescopes was the color fringes that were produced and which effectively blurred the image. The fringes which appeared around the image resulted because the degree of bending of light varies, increasing from red to violet. You recall that we discussed this earlier in Chapter 3 when talking about prisms. A lens functions as a double prism, the arrangement of the prisms varying as the shape of the lens varies.

Lenses used today are usually compound rather than simple. A simple lens is made of one kind of glass. A compound lens is a fusion of two or more different kinds of glass (glass made from different materials), each of which bends light to a different degree. Such a lens brings both long and short wavelengths to the same focus, thus avoiding color fringes.

Compound lenses were not available to early telescopemakers. For almost a hundred years they had to contend with blurry images. Toward the end of the seventeenth century Isaac Newton solved the problem—not with a compound lens but with a different kind of telescope, one that gathered light with a mirror instead of a lens. It was called a reflecting telescope. In place of the objective lens that refracted light, Newton used a mirror that reflected light. The mirror was curved so that the light gathered by it was brought to a sharp focus. The image was then magnified by using a viewing lens.

It became possible to build larger and larger telescopes because mirrors can be made much larger than can lenses. In England, William Herschel perfected the art of telescopemaking. After dozens of failures he found the way to make smooth, shiny curved mirrors, and also lenses that were smooth and perfectly curved. A hundred years after Newton had invented the reflecting telescope, man was at last able to

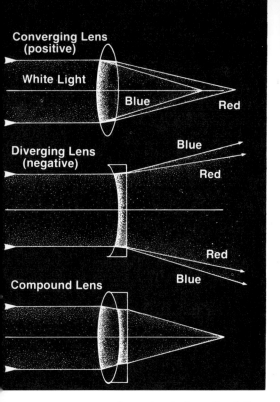 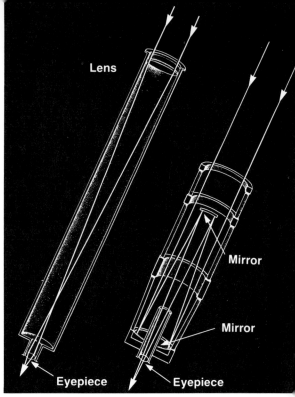

In a simple lens (top) the various wavelengths cannot be focused tightly; some are bent more than others. A diverging lens (middle) does not focus wavelengths. A compound lens (bottom) combines lenses so all wavelengths are brought to the same focus.

In these cross-sections one can see that in a refractor telescope (left) light is gathered by a lens system. In a reflector telescope (right) light is gathered by a system of mirrors.

see distant objects clearly, and he was able to gather together light from extremely dim sources. Today's large reflecting telescopes can see objects that are 10 million times dimmer than those that can be seen by the eye alone.

While telescopes made it possible to see dim objects, viewing was still instantaneous. That is, the human eye cannot store up light. If something cannot be seen because there is not sufficient light, no amount of looking at it will make the object visible. However, photographic film can "see" such objects because the effect of light on the film is cumulative, as during a time exposure. Light so dim that it cannot be seen by the eye, can be made seeable by collecting the light in a camera.

The telescope is a light gatherer; light from a large area (the cross section of the objective) is brought to a sharp focus. A camera is a light-deposit box; light gathered by the telescope registers on the film, producing a record of the observation. In the last half of the nineteenth century the camera emerged as a primary tool for the astronomer. Actually, large telescopes are specialized cameras; they are used to gather light, store it up on film, and so make a record of the observation.

SPECTROSCOPY

In the 1860s Kirchhoff, whom we mentioned earlier, and Robert Wilhelm von Bunsen (1811–1899), a German chemist, measured accurately the various wavelengths of which white light is composed, and established the field of spectroscopy. The light of a star, for example, was separated into its many parts. Each furnished information about the star—its temperature, what it was made of, the manner in which it was moving, whether or not it had a magnetic field, if the star was calm or explosive.

A continuous spectrum gives very little information about the object that is producing it. Actually, the continuous spectrum of one substance appears very much like that of another substance.

Bright emission lines and their counterparts, the dark absorption lines, are signatures of the stars—or of whatever else is producing them. Analysis of the lines enables one to determine the elements of which the stars are made. The procedure is straightforward. A bright-line spectrum of iron, let us say, is made in a laboratory by registering the spectrum of iron vapor on photographic film. In a similar manner the dark-line spectrum of a star is registered on film through a combination of telescope, spectroscope, and camera. The two spectra are then compared. If iron lines match up with lines in the spectrum of the star, then iron must exist in the star.

Spectrograms of the Sun and other stars, of comets, meteors, and planets, reveal that they all contain the same materials. The amounts vary a great deal; but the universe appears to be made of the same basic elements that we recognize here on our own planet.

Had spectral lines revealed nothing more than the substances producing them, they would still have been of immense value in probing the universe. However, the lines disclose much more. One of the types of information contained in them was discovered in 1842 by the Austrian scientist Christian Doppler (1803–1853).

Doppler did most of his work with sound, but he surmised that many of his findings would apply equally to light. Armand Fizeau, the French scientist who studied the speed of light, was to demonstrate a few years later that light waves behave as do sound waves in certain respects.

When a sound-producer—a jet plane or the horn of a car, for example—is moving toward you the sound increases in pitch. (It gets louder, too, but we are not concerned with changes in loudness.) As the sound-producer goes away from you, the sound drops in pitch.

Let's consider the approaching jet. It produces a sound wave. The next moment the jet is a bit closer to you, still

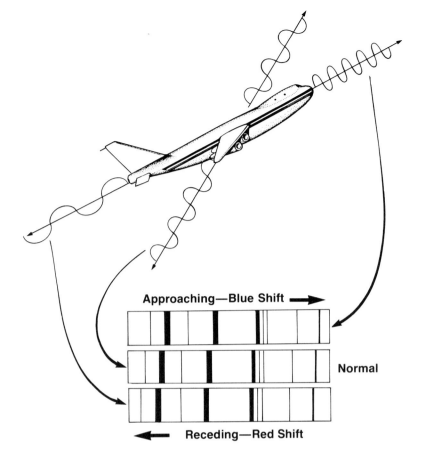

When approaching, waves of sound and radiation are shifted toward the blue. When receding, the shift is toward the red.

producing a sound wave. The wave produced at that moment is closer to the first one than it would be if the plane were not moving. The wave produced at the next moment is even closer. And so it goes, each successive wave a bit closer than the previous one. The closer together the waves are, the higher the pitch. When the jet is going away from you, the opposite reasoning applies. Each wave is a bit farther apart than the preceding one. Wavelength increases, and so the pitch drops.

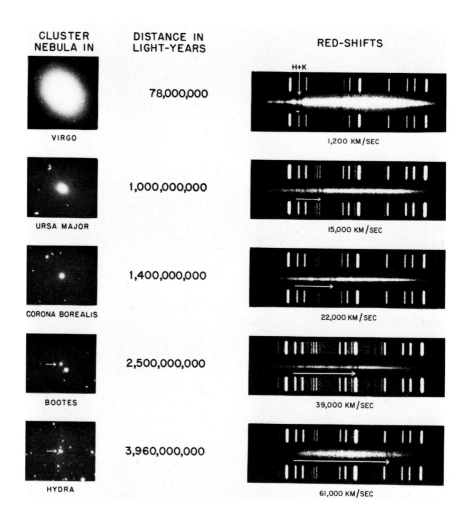

Distances to galaxies can be determined by measuring the amount of red shift. (Hale Observatories)

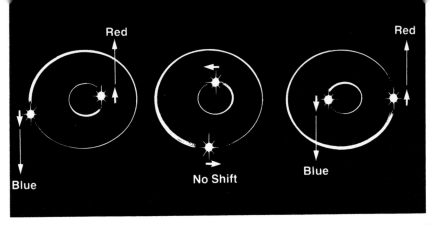

In double stars, blue and red shifts alternate as a star approaches and recedes.

Fizeau applied this reasoning to the behavior of light. He showed that when a light-producer (a star, for example) is moving away from us, the light is shifted toward the red (wavelength is increased). Should the star be approaching, the line shift is toward the blue (wavelength is shortened). The greater the shift, the greater the velocity of the object. The same observations could be made even though if the object producing the light is not moving; in that case the observed shifts would be due to motion of the observer.

These relationships are of immense value in astronomy. They enable determination of the velocities of stars and galaxies. Since the great majority of alaxies show red shifts, they are moving apart; the universe is expanding.

Furthermore, those galaxies that are farther away from us show a greater red shift than those that are nearer. This is exactly what astronomers would expect to find since the universe is expanding and since, as seems very likely, it all began billions of years ago with a great explosion of primeval matter.

Doppler shifts of light (or the Doppler-Fizeau effect, as it is often called) also reveal to us stars that could not be observed otherwise. Many of the stars that seem to be single objects are actually two stars, one going around the other. However, because the stars are so far away or so close together, we cannot observe them as doubles. Spectroscopically, however, their motions are revealed: shifts of the lines toward the blue as a star approaches, and toward the red as it recedes.

Because of their great distances, stars appear as points of light. However, the Sun is close enough so we see it as a disk. The edges of the disk can be observed. It has been found that one edge shows a blue shift, while the opposite edge exhibits a red shift. In other words, the Sun is rotating. The amount of shift indicates a velocity of about 7 000 kilometers an hour —sufficient for a complete rotation of the equatorial region in about 26 days.

SPECTRA AND MAGNETIC FIELDS

In 1896, Pieter Zeeman (1865–1943), a Dutch physicist, discovered that the spectral lines of light are split when passed through a magnetic field. Some of the lines are shifted toward the blue, others toward the red. The observation has been useful in many ways. For example, when light from a Sunspot is passed through a spectroscope splitting appears at the sunspot. The lines above and below the spot are normal, indicating that a magnetic field is associated with the sunspot. Sunspots are magnetic disturbances in the solar surface. We know this because of analysis of spectral lines of the Sun.

SPECTRA AND ROTATION OF THE GALAXY

Our galaxy is made of stars, dust, and gases, separated from each other but all moving around the central part of the galaxy. Those closest to the center move the fastest.

As we observe stars in our area of the galaxy, we find that

when we look at the stars "ahead" of us and closer to the center, the lines are shifted toward the red—these stars are moving away from us. The stars "behind" us and closer to the center show a blue shift—they are approaching us, catching up. When we look at stars that are farther from the center of the galaxy than we are, the observations are reversed. Those stars that are head of us show a blue shift—we are catching up to them. Stars behind us show red shifts—we are pulling away from them.

These observations, together with other lines of reasoning, reveal that we are located about 30 000 light years from the center of the galaxy. Our velocity around the center is about 250 km/sec, and it takes about 250 million years to complete a rotation—a cosmic year.

Analysis of the spectrum of a star reveals much more. It tells us whether a star is expanding (red shift) or collapsing (blue shift) at a given moment. If there's a succession of blue to red we know that the star is pulsating. A star ejecting gases at high velocities will also show a red shift. Some stars have very hot extensive atmospheres; they have a dark-line spectrum surrounded by bright lines.

We have discussed several of the ways in which the telltale emission and absorption lines can be read. The spectrograph has become an essential instrument for the astronomer, along with the telescope and camera.

Not only do these lines tell us about the macrouniverse (the big things), but they also reveal secrets of the microuniverse (the small things).

SPECTRA AND CHEMICAL STRUCTURE

A neutral atom is one in which the number of electrons and protons is the same. When an electron is removed, the atom becomes ionized. The number of different ions that the atom may become is limited only by the number of electrons. Each of the ions produces a distinctive spectrum. By analyzing the spectrum of a given element, one can determine if the atom is neutral or ionized and, if it is ionized, the degree of ionization.

Also, should the atom of the element be combined with another atom (H_2, for example), a molecule is created. The spectrum of molecular hydrogen will be different from the spectrum of atomic hydrogen.

In a spectrochemical laboratory the composition of a substance can often be determined in a few minutes. And the percentage of each element making up the substance will be indicated. The number of lines in the spectrum may be in the thousands and they may appear to be completely undecipherable. However, instruments and techniques for analysis have been developed. The time needed for a spectral analysis of a substance is most often much less than the time needed to analyze a substance chemically. Also, the accuracy of the results is very high.

ATOMIC STRUCTURE

At first glance the spectra of elements appear to have no structure. However, you recall that in 1885 J. J. Balmer discovered that a mathematical relationship exists between the lines of a substance. Later on, other relationships were found by other investigators. For example, it was found that there are several series of double lines (lines close together) for the alkaline metals—lithium, sodium, potassium. Also, there are triple lines for the magnesium-calcium-zinc group. Further investigation indicated that such peculiarities are due to the positions of electrons in the atoms—their energy levels.

From the spectra they produce, atoms and molecules can be identified. Also, scientists are able to deduce the internal arrangement of the electrons, and whether the atom has lost or gained electrons.

During a solar eclipse in 1868 the French astronomer J. Jannsen was studying the chromosphere, a lower layer of the Sun's atmosphere that is highly visible at totality. In the spectrum obtained at that time Jannsen found a bright yellow line that had never been observed before. At first it was thought that the line might be associated with sodium. Perhaps it was a line that sodium produced only when subject to the unusual conditions that exist on a star such as the Sun.

However, shortly after the discovery, the English scientists J. N. Lockyer and E. Frankland announced that the line observed by the French astronomer was produced by an element, one that had not been found on earth. They called the element helium, after *Helios,* the Greek word for the Sun. About a decade later helium was found on the earth. The story of helium is a dramatic example of the effectiveness of spectroscopy in analyzing the materials present in a light-producer.

10 The Radio Spectrum

Visible light is made of short-wave radiation—around 50 millionths of a centimeter. Somewhat longer waves are classified as infrared. When wavelengths become even longer, when they increase to a few millimeters and extend to several hundred meters, the radiation is classified as radio waves.

You are surrounded by them. You are not aware of their presence because no part of the human body is sensitive to radio waves, in the way eyes are sensitive to light. However, you know that the waves exist, for a radio receiver will pick them up almost everywhere on the surface of the earth. (These are waves sent out by manmade transmitters.)

In addition to manmade radio waves, you are surrounded by waves that are generated by the Sun, by planets, galaxies, gas clouds, and molecules that exist in space between the stars. The total energy in the waves is very small—so small that an ordinary receiver cannot pick them up. However, radiotelescopes can. These are large antennas connected to supersensitive receiving sets.

Jansky's antenna for receiving radio waves. (Bell Laboratories)

Reber's radio telescope. (National Radio Astronomy Observatory)

The history of optical telescopes began in the early 1600s, almost four hundred years ago. The existence of celestial radio waves was not even suspected until the early 1930s, only a few decades ago. Not until the last half of this century did the new science of radioastronomy really make strides. Its start was quite accidental.

In 1932, Bell Telephone Laboratories was trying to find out why long-distance radiotelephone communication was interrupted at regular intervals by what appeared to be static. Karl Jansky (1905-1950), an American radio engineer, was assigned to find some answers. To do so he built a large radioantenna. It was mounted on wheels that rolled on a circular track, making it possible to turn the antenna all the way around.

Jansky found that the disruption occurred when the antenna was pointed at the Sagittarius region of the sky, the location toward the center of the galaxy. The static occurred every 23 hours 56 minutes—exactly the time required for the Sagittarius region to reappear at the same location. The radio waves interrupting radiotelephone links were apparently coming either from Sagittarius or from the center of the galaxy. Without any intent to do so, Jansky had discovered celestial radio waves.

In December 1932, Jansky published a report of his work in a radio engineering journal, a magazine that astronomers of that day did not read. Grote Reber (1911-), an American engineer who was also an amateur astronomer, read the report. He was determined to explore celestial radio waves, and he set about designing a radiotelescope. In the mid-1930s he built the first radiotelescope reflector, or dish, as this type of radiotelescope is called. It could be swung all the way around, just as Jansky's telescope. Also, the dish could be moved up and down from horizon to zenith. With his reflector, which was 10 meters across, Reber discovered that the Sun produces radio waves, and he also discovered that radio waves come from other parts of the galaxy in addition to the center. He produced the first radio map of the galaxy. Looking back, Karl Jansky must be credited with the discovery of celestial radio waves and the first application of a radioantenna to sky scanning; and Grote Reber with the construction of the first dish-type radiotelescope.

Today the largest dishes, or parabolic reflecting telescopes, are about 92 meters across. By combining several smaller telescopes the effective size of a radiotelescope can be made much larger. The dish of a telescope is a collector of radio energy. The collector reflects the waves to an antenna mounted at the focus; energy from the entire surface of the bowl is brought to a point. The radio waves are fed to a detector-amplifier, which separates the waves (tunes some in and others out) and strengthens them. The radioastronomer can listen to the radio signal—although there is little to be gained from listening alone, for the sounds are hard to discern from background sounds ("noise," as it is called). More important, the signal is recorded onto paper tape, providing a visual record. For quick recall, convenience, and ready reference, however, a computer is required. The signals are fed into a magnetic tape recorder, which in turn converts the signal to a series of numbers that are printed on computer cards. This record of the observation can be stored easily; it can be recalled with ease, and organized or processed in a variety of ways depending upon the goals of the astronomer.

Radiotelescopes need not be dishes. They are often extensive arrays of antennas, sometimes covering several acres. They can therefore pick up very weak signals, just as very large TV antennas must be used in places where manmade signals are weak.

THE RADIO SPECTRUM

The radio waves received by Jansky and Reber were continuous; the spectrum they produced was a broad band rather than one of discrete lines. You could compare it with the band of colors produced by hot solids or by gases when they are packed together tightly.

In 1944 the Dutch astronomer H. C. van de Hulst suggested that hydrogen atoms in space should act as tiny radio transmitters. And, rather than producing a continuous spectrum, they should produce a single wavelength, a "bright line" located at 21 centimeters. In 1951 radiotelescopes picked up the predicted 21-centimeter line. You recall that bright emission lines of light radiation are the signatures of the elements. The bright emission lines of radio radiation are equally informative. They tell the astronomer the temperature of objects, they disclose the structure of molecules, and they reveal the activity of electrons of the atoms.

Bright radio lines ushered in the age of radioastronomy. From the early 1950s radioastronomy has grown into one of the most exciting branches of astronomy. There are many reasons for this. Radio telescopes can "see" into regions of space that optical telescopes cannot penetrate. Radio waves come through our atmosphere without interference, unlike other parts of the spectrum—many of the infrared waves are removed by water droplets; ultraviolet rays are largely removed by oxygen atoms; and visible waves are obscured by clouds. The X rays and gamma rays, too, are dispersed by atoms and molecules in the atmosphere. Radiotelescopes are just as useful in daytime as they are during the night; daylight does not interfere with the reception of radio waves.

PRODUCING THE RADIO SPECTRUM

The visible spectrum, the continuum and also the bright lines, are produced when atoms are agitated. The radio spectrum is produced in a similar manner, by electrons when they are agitated (actually by the collisions, or near collisions, of electrons in many cases). Unless the temperature of a substance is absolute zero—0°K—its electrons are moving. Since absolute zero probably does not exist anywhere, every object in the universe is producing radio waves. The waves may not be discernible by our equipment; nevertheless they exist.

Waves produced by the collisions of electrons are said to result from thermal emission ("thermal" meaning "heat"). The hotter an object is, the more intense will be the waves. This means that the temperature of an object, say, a planet or a galaxy, can be determined by analysis of the radio waves that it produces.

When the electrons are moving at ultrahigh speeds, close to the speed of light (300 000 km/sec) they produce radio waves by what is called synchrotron emission. The electrons spiral along magnetic lines of force in space much the same as they move along the lines of force in earth-based synchrotrons—nuclear accelerators. Many of the radio waves received from beyond the solar system are produced in this manner.

Sharply defined radio waves—waves at specific wavelengths, which produce a "bright-line" radio spectrum—are produced when electrons in the atoms change levels—when the electrons change orbits, or when they change direction of spin. The protons and electrons of atoms may be compared with the Sun and earth. They revolve around each other and, just as the Sun and earth rotate, so also do the protons and electrons. The 21-centimeter line of hydrogen is produced when the electron of the atom flips over; effectively, it spins in the opposite direction. In a given undisturbed hydrogen atom the flipover happens very seldom—once in about 11 million years. However, because of collisions between atoms which stimulate spin reversals, the flipovers occur once in about 400 years. Even though the chance of the change occurring in any given instant in any random atom is small, there are so many hydrogen atoms in space, and space itself is so vast, that the 21-centimeter line exists more or less continuously and intensely. Scanning space with telescopes tuned to pick up the 21-centimeter wavelength has enabled astronomers to "see" a major part of the galaxy.

MAPPING THE GALAXY

Scores of discoveries were made in astronomy during the early years of the seventeenth century. The newly discovered optical telescope was so much more powerful a tool than the eye alone that every time something was observed, a discovery was made: the craters of the Moon, the phasing of Venus, the rings of Saturn, the satellites of Jupiter.

In our present century astronomers are in a similar situation. Radiotelescopes are new tools astronomers are learning how to use. In only a few decades radioastronomy has already given us a great deal of information about the universe.

Optical telescopes gave us views of the stars, dust, and gases in our galaxy. In the 1920s there was considerable

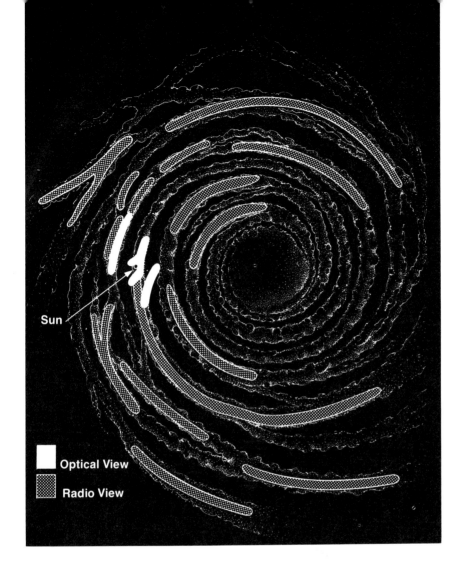

controversy about what the observations meant. Some astronomers believed that they indicated the Sun was at the center of the galaxy, much as William Herschel had suggested more than a century earlier. Others believed the observations indicated that the Sun was removed from the center, and the galaxy itself was disk-shaped. Additional observations led to the belief that our galaxy is about 100,000 light years across and about a thousand light years thick. The Sun is about 30,000 light years from the center. It is located in a spiral arm and is moving around the center of the galaxy at a speed of some 250 km/sec.

The optical view of the galaxy has revealed a great deal. Still, the optical perspective is restricted. Light from distant stars and clusters is filtered out by clouds of gas and dust. Even though astronomers have seen millions of stars, and have catalogued hundreds of thousands of them, most of the stars of the galaxy have escaped their perusal. Judging by the mass of the galaxy there are, conservatively, 100 billion stars in it. Even if a million stars had been catalogued (which they have not), this would be a small fraction of the total—only about one hundred-thousandth. Some feeling for the comparative areas of the whole galaxy and the restricted portion open to visual observation can be gained from the illustration on page 91.

To define the shape of the galaxy more accurately and to determine the shape and dimensions of the arms that appear to spiral out from the central region, radioastronomy is required. By tuning to the 21-centimeter line of hydrogen, astronomers can work out its distribution. When the information is presented pictorially, a map such as that shown on page 91 emerges. But that map needs correcting and enhancing. This is one of the areas that radioastronomers continue to investigate, always hoping to be able to construct more and more detailed representations of the galaxy in which we live.

TEMPERATURE OF THE PLANETS

You recall that unless its temperature is absolute zero, a body will give off radio waves in a way determined only by its temperature. Therefore if we can detect radio waves from a planet, an analysis of those waves should tell us the temperature of the planet. And it does. Temperatures of the planets are given in the table below.

PLANET	TEMPERATURE (°C)	
	SURFACE	CLOUDS
Mercury	350 (day)	
	−170 (night)	
Venus	480	−33
Earth	22	
Mars	−23	
Jupiter		−150
Saturn		−180
Uranus		−210
Neptune		−220
Pluto		−230 (?)

The radio spectra of most of the planets are continuous. However, those of Venus and Jupiter are exceptions. They indicate that unusual conditions exist on both of the planets.

Venus

From visual observations it has been dete Venus is covered with a blanket of opaque cloud since radio waves can penetrate clouds, one wou the radio waves we receive would be emitted by the surface. The temperature of the clouds of Venus, as measured by other techniques, is about −40 degrees Celsius. It was a great surprise when the surface temperature, as measured by the radiotelescope, turned out to be about 480°C—far greater than the sea-level temperature of boiling water.

The clouds of Venus apparently act as a roof, one that

permits radiation from space to pass through but which turns back radiation that emanates from the surface. The process is similar to that of a greenhouse—sunlight can enter through the glass; but heat (long-wave radiation) cannot escape. The clouds are a blanket that holds in heat, thus allowing it to build up.

Jupiter

Jupiter is a cold planet—so cold that its atmosphere contains frozen crystals of ammonia and methane. Infrared measurements reveal a temperature of about $-150°C$. However, when the planet was scanned at various radio frequencies, temperatures ranging from 400 to 50 000°C were indicated. Obviously something was wrong; the result could not be produced by ordinary agitation of the atoms and electrons in the atmosphere of the planet. If this were so, the temperature would be uniform. The radio signal had to originate in some other fashion.

Observers reasoned that the variety of radio temperatures could be caused by high-speed electrons accelerated in a powerful magnetic field. Earth is surrounded by such electrons trapped in doughnut-shaped belts—the Van Allen radiation belts. Perhaps similar regions surrounded Jupiter.

Observations were made with sensitive interferometers. One drawback of a radiotelescope is its inability to focus tightly, to "see" the details of a radio wave transmitter, or to narrow the field of view. In order to do this a telescope several miles across would be needed—obviously quite impossible. But to get the same result two telescopes can be used. They are placed far apart and joined together by cables so the signals they pick up go to a single receiver. Such an arrangement, called an interferometer, was aimed at Jupiter.

It revealed that the signals were not coming from the planet itself but from a region beyond the surface—apparently from radiation belts. The radio temperatures indicated that the particles were intensive and the magnetic field powerful.

In addition to the signals mentioned above, Jupiter emits bursts of radio energy at regular intervals. The cause of these bursts has not been determined. However, some observers believe they are related in some way to the motions of one or another of Jupiter's satellites, perhaps Io, the innermost of the four that were discovered by Galileo. Conceivably, when Io is in a particular position relative to the earth a sort of window is produced, one that permits surges of radio energy to escape from Jupiter's radiation belts. As equipment becomes more sensitive, we may find whether an explanation somewhat like this is correct, or if the phenomenon is caused by some entirely different condition.

PULSARS (NEUTRON STARS)

In the late 1960s an exciting discovery was made: neutron stars were found by analyzing radio waves that they emitted. Astronomers at Cambridge, England, were investigating the "twinkling" of radio waves. (Stars twinkle because of interference of the light by earth's atmosphere. In a similar fashion radio sources twinkle, or scintillate. In this case the interference is caused by irregularities in the space around the solar system—interplanetary space.) Jocelyn Bell, one of the researchers, noticed that there was one interference that occurred regularly and that could not be due to the Sun or anything else in the solar system. The pulsations were extremely regular and very rapid—every 1.33 seconds.

Because of the nature of the signal, it was at first suspected that it might be some sort of celestial communication system.

The source, whatever it might be, was referred to as "LGM" (Little Green Men). However, later investigations indicated that the Cambridge astronomers had detected not a radio pulse generated by civilized creatures but a rapidly pulsating star—a pulsar. Theorists had long claimed that there were such things—neutron stars, they called them. In 1967 they were proved correct. Since the discovery of the first one, more than a hundred pulsars have been identified. In most cases the only way we know of their presence is by the radio waves they give off. However, an exception is the pulsar associated with the Crab Nebula.

In 1054, Chinese astronomers reported that suddenly a bright star had appeared in the constellation of Taurus, in a region where previously there had been no star. For a time it was so bright the star could be seen in the daytime. As it dimmed, the star was observed for several weeks before it faded into the night sky.

When modern telescopes were turned to this part of the sky a tremendous mass of gases was seen—a formation called the Crab Nebula. The gases were in rapid motion, speeding away from the central region at a velocity of some 1 200 km/sec. The star seen by the Chinese was a supernova—a star that had exploded. Shells of gases were ejected by the star, and tremendous amounts of energy. The Chinese saw the light energy. Modern astronomers "see" the radio energy.

In 1968 astronomers at Green Bank, West Virginia, turned their telescopes to the Crab Nebula. As predicted, they detected strong radio waves coming from that direction. They focused in more tightly and found that the waves were coming from a narrow point, a star. Furthermore, the star was

Strong radio waves, and also X rays, apparently coming from the Crab Nebula in Taurus, have been detected. (Hale Observatories)

pulsating rapidly. The star at the center of the gases was a pulsar. Optical astronomers turned their telescopes to the same location, and they were able to see a pulsar for the first time.

Astronomers had long suspected that certain stars blew apart as they moved along their life cycles. They believed that a remnant of the explosion would be a star made up mostly of neutrons, and that the star would rotate rapidly. Radioastronomy had produced the facts to back up the theory.

MOLECULES IN SPACE

In a laboratory, radio waves are aimed at a gas. Using a microwave spectroscope the scientist determines which wavelengths the gas absorbs. He then knows that under certain circumstances that gas would emit those particular wavelengths.

In the early 1960s a substance known as hydroxyl (OH) had been studied in this way, so astronomers knew which wavelengths would identify OH. The wavelengths showed up as four distinct lines on the radio spectrum. At the Massachusetts Institute of Technology a radiotelescope was pointed toward Cassiopeia, and soon after one of the lines of OH was found. After a time more of the lines were identified. By 1965 other researchers had found the lines also, and not only in the Cassiopeia region. Thus began a search not only for OH but for other molecules in interstellar space.

By the early 1970s dozens of substances had been found: water, ammonia, hydrogen sulfide, alcohol, formic acid, among them. In some cases the molecules were emitting radio waves. In more cases the molecules were absorbing certain wavelengths that were being produced by gas clouds located far beyond the molecules; the molecules filtered out certain wavelengths. The list shows the molecules that have been identified and the frequency of the lines by which this was done.

Astrobiologists and astrochemists continue to explore these exciting discoveries. Just what the presence of molecules in space implies is not known. Some believe they strengthen the argument that life is not peculiar to earth. Indeed the existence of life may be a common condition in the universe. The raw materials for it seem to be abundant.

THE BIG BANG—3° RADIATION

Radio waves from outer space have provided a partial answer to one of the basic questions of astronomy—what kind of universe are we living in? At present there are three possibilities: (1) the universe is steady, that is, as it expands new matter is created, filling in the space that the expansion has created; (2) the universe is pulsating—it expands and then contracts, only to expand once again; (3) the universe is

Some 15 billion years ago a "super-atom" exploded, producing the universe of stars and galaxies.

continually expanding, having begun when a single superatom containing the matter of the universe exploded.

Of the three ideas, the last, which is called the Big Bang Theory, is most widely accepted. If we were able to look far enough into space, which is the same as looking backward in time, we should be able to see the universe as it was in the beginning. Or, at least, some remnants of that far-off time might be seen.

Theoretical astronomers have predicted that if there had been an explosion billions of years ago, cooling would have been going on ever since the event. The amount of cooling would have reduced the original temperature of perhaps a billion degrees to 3°K—3° above absolute zero.

Radioastronomers believed that if they could aim a very sensitive receiver at a blank part of the sky, a region that appeared to be empty, it might be possible to determine whether or not the theorists were correct. This was done in the early 1970s. Two scientists at Bell Telephone Laboratories (the same place where Karl Jansky had discovered cosmic radio waves) picked up radio signals from "empty" space. After sorting out all known causes for the signals, there was still left a signal of 3° they could not account for. Since that first experiment others have been carried out. They always produce the same result—3° radiation.

Space is not absolutely cold. The temperature of the universe appears to be 3°K. It is the exact temperature the universe should be if it all began some 13 billion years ago, with a Big Bang.

We've discussed only a few of the exciting discoveries that exploration of the radio spectrum has made possible. This is a young science, and no doubt as it matures additional equally impressive discoveries will be made, revealing more and more about the universe.

The universe abounds with light energy, and we are readily aware of it. And it is also filled with radio energy. All we need is the proper equipment to be aware of its presence, and sufficient ingenuity to figure out what it means. With appropriate equipment we can also detect many other kinds of energy (radiation) which hold fascinating secrets.

11 The Infrared Spectrum

Unless the temperature is absolute zero, all objects give off radiant heat. The amount may be very small, so small it is not detectable by the most sensitive instruments. Even large amounts of radiation cannot always be detected easily. The darkest theater, for example, is full of infrared photons. They are given off by the warm bodies in the audience. Their wavelengths would be about twenty times longer than that of the green photon of sunlight, and their energy about twenty times less. The eye cannot detect these photons, for vision sensitivity drops about a million-million times when the wavelength of the photons is only double that of green light. Infrared (below the red) radiation has a wavelength just a bit longer than red light (about 700 nanometers) and extending to some 3 million nanometers (3 millimeters), the region that can be detected by radio receivers. (A nanometer, you recall, is one millionth of a millimeter.)

William Herschel, the musician-turned-astronomer, who

was also an outstanding telescope-maker, is credited with discovering infrared radiation. About 1800 he observed the Sun through various filters, or, as he called them, "differently coloured darkening glasses." Reporting on his observations, he said, "What appeared remarkable was that when I used some of them, I felt a sensation of heat, though I had but little light; while others gave me much light, with scarce any sensation of heat."

To investigate these observations, Herschel directed sunlight through a glass prism to produce a spectrum of the sunlight. He blackened the bulb of a thermometer, then moved the thermometer through the various sections of the spectrum, from violet to red. He discovered that the temperature increased steadily as the thermometer was moved toward the red end of the spectrum. Even more startling, the temperature remained high for some distance beyond the visible red. Later on, Herschel showed that radiation below the red was produced by warm objects other than the Sun. Also, this invisible radiation behaved in many ways much the same as does visible light. Some fifty years later, more precise experiments conducted by the Italian physicist Macedonio Melloni (1798–1854) were to confirm Herschel's discoveries.

Though it was sufficient for Herschel's pioneering experiments, glass is a poor conductor of infrared. Melloni used crystals of sodium chloride (table salt), which were much more transparent to infrared radiation. He made salt lenses and prisms and showed that infared can be reflected and can be spread into a spectrum, much as visible light can. Also, the infrared spectrum could reveal information about the body that was producing it. The problem became one of designing devices sensitive enough to detect the radiation.

INFRARED DETECTORS

Herschel's detector was his own skin. Infrared photons fell upon him and were converted into sensible heat—heat that he could feel.

One of today's simplest devices for detecting the presence of infrared radiation is an ordinary camera loaded with infrared film. The film is sensitive to what is called the near infrared. That's radiation in the region between about 700 and 900 nanometers. Infrared radiation extends to wavelengths of around 3 million nanometers before it merges into radio waves. But wavelengths beyond about 900 nanometers are difficult to detect photographically even when stray photons are controlled. This is done by chilling the equipment to supercold temperatures.

To see some interesting effects of infrared radiation, load your 35-mm camera with high speed infrared film. Detailed exposure information will be found in the folder that accompanies the film. If you photograph a landscape in summer when there is heavy foliage, you'll find that clouds appear very white against a dark sky, leaves of trees appear light, water appears dark.

Interesting effects are obtained by taking pictures in complete darkness. The object to be photographed can be "illuminated" by a flatiron. The infrared radiation given off by the iron reflects from the object being photographed—books, bookends, a vase, or whatever.

Infrared radiation penetrates haze effectively. Try taking a picture of a hazy landscape. You'll find that the infrared film reveals distant trees and other objects as though the haze were not present. If you wish, you can also experiment with color photography, for there are films that produce unusual

An aerial view of the Detroit, Michigan–Windsor, Ontario area (top), and the same area photographed in infrared light (bottom). In the bottom photo the details, especially the island, are sharper. (NASA)

pictures in which foliage appears red and plowed land seems to be snow-covered.

These experiments with infrared photography produce unique artistic effects. But the detection of infrared radiation has also become a valuable tool that provides scientists with various kinds of information. For example, in aerial photos of orchards, trees that are diseased are blue-green, while healthy trees are red; under infrared certain types of forgeries of documents and paintings show up clearly. In medicine the infrared camera reveals tumors under the skin, blood circulation patterns, arterial blockages. There are multitudes of other uses, and in many different branches of study: infrared reveals many of the causes of water pollution, the kinds of rock in a given formation. Among the most important information revealed by infrared is that which is related to the field of astronomy.

INFRARED IN ASTRONOMY

Astronomers use film, such as that mentioned above, as a detector of infrared. They also use electronic devices that are affected by radiation. One of the earliest is called a thermocouple. It was first used in astronomy by William Parsons, third Earl of Rosse (1800–1867). A little over a hundred years ago he used his famous telescope, called the "Leviathan of Parsonstown" because of its great size, to explore the infrared radiation between 800 and 1 400 nanometers given off by the Moon. A thermocouple consists of wires of two different metals twisted together. The other ends of the wires are connected to an indicator. When infrared radiation is focused on the junction of the wires, a small current of electricity is generated, the amount being related to the amount of radiation that is received. Another detector that operates on the same principle is the thermopile—several thermocouples wired together so they act as a single unit. Other detectors operate because the resistance in an electric circuit is changed when exposed to infrared. Even more sensitive detectors use crystals. Infrared photons striking the crystals cause them to give off electrons, that is, generate an electric current. Another type of detector does not depend upon electricity. A gas is heated by the radiation. This pushes on a sensitive mirror, which causes a beam of light to be deflected. The deflection of the beam, which is related to the amount of radiation, can be measured.

Infrared detection from a specific object is difficult because the detector is affected by all infrared radiation, not just the radiation coming from the object being studied. It is as though a visible-light astronomer were observing the stars in full daylight with a luminous telescope—one that was giving off light. In order to reduce infrared radiation from the instrument itself, astronomers use liquid nitrogen to cool the detector to just a few degrees above absolute zero.

The entire infrared region of the spectrum is a broad band extending from about 700 to about 3 million nanometers. Beyond that it merges into radio waves. However, not all infrared radiation reaches earth's surface. Certain wavelengths are absorbed by water and other molecules in the atmosphere. Astronomers therefore concentrate on those particular wavelengths that come through "windows" in the atmosphere. Several regions of the spectrum are relatively unexplored, including the entire band beyond about 22 000 nanometers.

Nevertheless the radiation received on earth has supplied

new knowledge; it has revealed temperatures of the planets (see chart on page 93), the rate of cooling of the Moon during an eclipse (from about 400°K to 150°K in just an hour), and much more.

THE SKY IN INFRARED

Most of the stars are not visible. They are cool red dwarfs and dark companions of much hotter stars, ones that give off visible light. Years ago astronomers believed that even more matter than had been suspected might be incorporated into stars too cool to produce visible radiation. Also, there might be vast clouds of gases just forming into stars—protostars, as they are called. The gases would be cool, their temperature only a few degrees Kelvin.

Surveys made of the whole sky at various wavelengths reveal that there are such concentrations. They have been noted in several locations, and at places where no visible star exists. Infrared provides a probe enabling astronomers to "see" what is happening in stellar evolution long before the "star" can be detected by other means. Also, infrared enables astronomers to look into the center of the Milky Way Galaxy, a region that is obscured visually by the gases that lie between us and the center. Infrared penetrates, however, and reveals that the concentration of stars at the core of the galaxy may be 10 million times greater than it is in the region of the Sun. If we were located there, the star-filled sky would be a magnificent display some 40 000 times brighter than that which we now see.

Like radioastronomy, infrared astronomy is a new area of investigation. In the years ahead improved detectors will expand the region that can be explored. Also, equipment located in spacelabs out beyond the obscuring atmosphere will open up large bands of infrared that have been denied our observation. No doubt we'll gather more and more information about the universe—where and how it came to exist, its composition, and its life stages.

12 Ultraviolet, X Rays, and Gamma Radiation

Just beyond the spectrum of visible light (about 400 nanometers) is the region of ultraviolet radiation. It ranges from about 400 to 30 nanometers. Most of it does not reach earth's surface—only that between about 400 and 300 nanometers. The rest is filtered out by the upper layers of the atmosphere. Shortwave radiation is removed at about 240 kilometers above the surface by atomic oxygen (O) and nitrogen (N) and also by molecular nitrogen (N_2). Radiation that penetrates to about 100 kilometers is removed by molecular oxygen (O_2). Below that level, at about 50 kilometers, most of the remaining radiation is removed by ozone (O_3), a gas that is produced when ultraviolet radiation reacts with molecular oxygen. Only a small amount of the ultraviolet of space penetrates to ground level. This is fortunate since large amounts of ultraviolet radiation are destructive. Small amounts, however, are beneficial.

The ultraviolet radiation received on earth originates in the Sun. For collecting the radiation and separating out certain wavelengths, a spectroheliograph is used. With it one can pick up the radiation given off by only certain atoms—calcium in the Sun, for example, or hydrogen—and so produce a picture of the calcium Sun or the hydrogen Sun. In the ultraviolet range wavelengths can also be separated, and so pictures of the ultraviolet Sun at different wavelengths (energy levels) can be constructed. For such views the spectroheliograph must be carried aloft by a rocket, a satellite, or a space laboratory.

Several Orbiting Solar Observatories (OSOs) have been launched since the first in 1962. They have surveyed the ultraviolet radiation at various altitudes in the solar atmosphere. The illustration below shows a photograph of the hydrogen Sun. Notice that the surface is uneven—there are light patches (hot regions) as well as many dark patches (clouds of cooler gas). The drawing records a survey of the lower atmosphere made in ultraviolet radiation. The shading becomes darker as temperature drops; the hottest and brightest regions produce the greatest amounts of radiation. When similar surveys are made at other wavelengths, the distribution of energy at levels up to several thousand miles is revealed.

Left, the sun as it appears in hydrogen light.
Right, the sun in ultraviolet light. (Scientific American)

 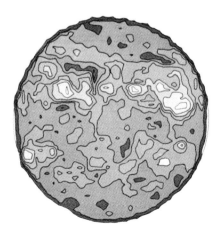

The OSOs and OAOs (Orbiting Astronomical Observatories), which followed, developed after the historic launching in 1946 of a modified German V-2 rocket. In October of that year a spectrograph was carried in the nose cone to an altitude of 80 kilometers. It photographed the solar spectrum down to a wavelength of 220 nanometers, well into the ultraviolet. This launching is regarded by many observers as the beginning of space astronomy.

You recall that electromagnetic radiation is produced when electrons change position in the atom. The changes can be brought about in many ways. One is thermally, by heat which accelerates subatomic particles, causing them to collide. The level of energy needed to produce ultraviolet radiation is higher than that needed to produce visible light. Also, as the temperature (the energy level) increases, the wavelength becomes shorter. By tuning the ultraviolet spectrograph to different wavelengths, temperatures at various locations in the solar atmosphere have been measured. Surprisingly, it was found that the temperature in the solar atmosphere becomes greater with altitude. Also, at the location where the chromosphere of the Sun (the lower atmosphere) becomes the corona (the portion visible during total eclipses) there is an abrupt jump from about 40 000°K to 500 000°K. The temperature at various altitudes is shown in the chart, together with the ions responsible for the detected ultraviolet radiation. The numbers in parentheses indicate the number of electrons removed to produce that particular ion. (An ion, you recall, is an atom that has gained, or lost, electrons.)

In 1968 telescopes were successfully placed in space when OAO–2 was launched. It also contained instruments that

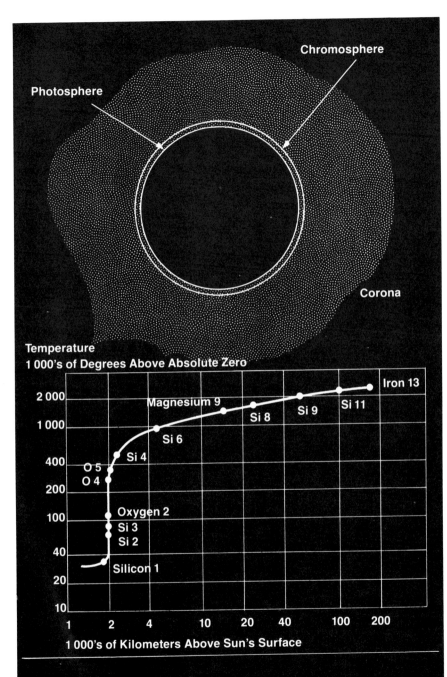

Drastic temperature changes occur in the solar atmosphere, the region above the chromosphere.

enabled the collection of ultraviolet radiation from the stars. Such data are important because extremely hot stars give off relatively little visible light. Most of the radiation is of shorter wavelength and not discernible except by ultraviolet-collecting techniques. Only by studying the ultraviolet spectrum can the astronomer determine the chemical composition of such stars and the rate at which they are evolving.

When the whole sky is scanned for ultraviolet radiation, the astronomer finds there are many strong sources. The hot, blue-white stars are prominent. So are pulsars (neutron stars) such as the one in the Crab Nebula; quasars, those puzzling objects that produce the energy of a galaxy of stars but which appear as single stars; and whole galaxies, such as the Andromeda Galaxy. This indicates that there are more young, hot energetic stars in a galaxy than had been suspected.

Ultraviolet radiation has revealed aspects of the universe not previously known to observers, aspects in the region of high-energy photons. Even more energetic radiation exists, however.

ROENTGEN AND X RAYS

Toward the end of the year 1895, Wilhelm Konrad Roentgen the German physicist, was experimenting with the conduction of electricity through gases. He removed much of the air from a glass tube and then passed an electric current through the tube. Nearby there happened to be some crystals of a barium salt (barium platinocyanide) spread upon a paper screen. When the current was turned on, Roentgen noticed that the screen fluoresced. When cardboard and glass were placed in front of the screen the fluorescence continued.

Roentgen investigated the phenomenon and found that the radiation (whatever it was) affected the photographic film much as light did. Certain materials, metals for the most part, cast shadows when placed in front of the barium screen. Since he did not understand the radiation, Roentgen called it X radiation or X rays. The name has persisted to the present day, even though we now understand more fully the nature of the radiation.

The radiation occupies that part of the electromagnetic spectrum that lies beyond ultraviolet. It is in the region having very short wavelengths, between 100 nanometers and 0.01 nanometer. As the wavelength of radiation decreases, energy level increases. Very-long-wave radio radiation contains extremely little energy; infrared radiation is more energetic; and X radiation is several thousand times more energetic than visible light, which is a step above infrared.

In the laboratory X rays are produced using many different kinds of materials. However, the principle is essentially the same in all cases. An X-ray tube consists of a cathode (negative) connection and an anode (positive) connection encased in a vacuum tube. In operation the cathode gives off electrons which strike the anode, causing the metal surface to give off X-ray photons. These photons travel through the glass tube and can then be utilized in many ways. They can penetrate steel to see if welds have been made evenly, or probe into crystals to determine their structure. And X rays are used extensively in medicine for diagnosis and for therapy.

NATURAL GENERATION OF X RAYS

When energetic particles, such as the electrons mentioned above, strike certain atoms, the atoms are disrupted. They are set out of balance. In regaining equilibrium, the atoms throw off photons. The photons may be at the energy level that produces visible light. Or they may be less energetic, and so radio waves are generated. When the bombarding particles are highly energetic, the photons given off will contain much more energy; they will be in the realm of X radiation, or even gamma rays.

In nature there are no X-ray tubes as such; however there are natural mechanisms by which X rays are produced. There would have to be, for there are are X-ray stars, X-ray hot spots, and an X-ray background permeating space, much as there is a background of radio energy.

The background radiation may be related to the presence of cosmic rays. These are highly energetic particles that may be remnants of the Big Bang, the cataclysmic explosion of the primeval "atom," or cosmic egg, from which came the stars and galaxies—the universe as we know it today. Our Milky Way Galaxy is disk-shaped—some 100 thousand light years across and 15 thousand light years thick at the center. A ball of low-density gases centered on the galactic nucleus and some 50 thousand light years across surrounds the galaxy. It is called the galactic halo. Photons of starlight-scattering cosmic ray electrons in this gaseous halo may produce the low-energy X rays that make the background radiation.

In certain very hot stars there are plasma layers. These are layers of atoms that have lost many if not all of their electrons —the atoms are highly ionized. In such gases high-energy electrons react with the ions in various ways. In one process, called bremsstrahlung, electrons are slowed down by atoms and photons are given off. When the reaction is at a high energy level the photons are in the X-ray region. (*Bremsstrah-*

Clusters of stars move around our galaxy. They form the halo of the galaxy.

lung is a German word meaning "deceleration radiation.") The electrons are not captured in the process.

Another reaction, one that was first observed in laboratories on earth, is called synchrotron radiation. (A synchrotron is a nuclear accelerator that speeds up electrons to velocities close to the speed of light.) In this process high-speed electrons spin around magnetic lines of force. Part of the energy contained in the electrons is given off in the form of high-energy photons. If the energy level is sufficient, the photons will be X rays.

The X-ray spectra produced by these processes can be differentiated from one another. Study of them enables scientists to determine many characteristics of the source, such as the strength of the magnetic field, the number of high-speed electrons, and the amount of energy they contain.

TELESCOPES AND X-RAY SOURCES

The atmosphere prevents the passage of X radiation because atoms and molecules in the air disperse the energy. Since this is so, instruments to detect X rays must be carried to an altitude of a few hundred kilometers by balloons, by rockets, or as part of the payloads of satellites and space laboratories. Glass lenses cannot focus X rays. But they can be deflected by metal plates. Therefore, plates in X-ray telescopes send the radiation through metal grids to a detector that converts the X rays into electrical pulses.

In 1962 the Italian physicist Bruno Rossi sent up a rocket that carried an X-ray telescope. He found that strong X rays were coming from the direction of Scorpius. The source has since been called Sco X–1. Since that historic flight several Explorer satellites have carried X-ray detectors. One of them, Explorer 42, launched from Kenya in Africa in 1970, was especially successful. It was named Uhuru, a Swahili word which means "freedom." For years Uhuru scanned the sky, and discovered over one hundred sources of X rays. The search goes on with other satellite-borne telescopes.

One of the sources pinpointed by Uhuru was located in Taurus. It turned out to be a neutron star, the remains of a star that had exploded several thousand years ago and was observed by the Chinese in 1054. It is a violently active, energetic star producing a wide array of photons, ranging from the radio band to the X-ray region. Some of the energy of this central star appears to be transmitted to gases surrounding the star and expanding away from it. As these gases collide with gas concentrations in space, additional photons are produced, some of these are X rays.

Radiation in the X-ray level has been detected coming from quasars, from entire galaxies, from pairs of stars rotating around one another, from black holes. These black holes are massive concentrations of matter in a small volume. Indeed, if a star such as the Sun were so concentrated, all its mass would be contained in a volume only a few miles across. The density in such an object is millions of tons per cubic centimeter. Gravitation is so great that nothing can escape

A black hole (bottom right) may be a companion of a very large, bright star. Because of its fantastic gravitation, the hole pulls gases out of the star.

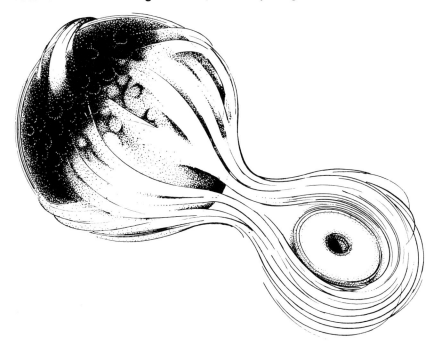

from a black hole—not radio waves, light, or X rays. Black holes may be one of two stars that are rotating around each other. Matter from one of the stars is pulled into the black hole. As the matter is pulled in, it releases some of its energy in an area within about 200 kilometers of the center of the black hole. Some of this energy will be in the X-ray region. A typical example of such a system is Cygnus X-1, the X-ray source in the constellation of Cygnus.

GAMMA RADIATION

At the frontier of the electromagnetic spectrum lie the gamma rays, the most energetic radiation. Far X rays blend into gamma radiation.

Gamma rays are produced when high-energy particles crash into each other, or into an atomic nucleus. The high-energy photons probably arise from violent nuclear processes and explosions such as those that occur in supernovas, and also as components of cosmic rays. Some researchers believe that gamma rays originate out beyond our galaxy in intergalactic space where cosmic ray remnants of the Big Bang still exist.

Telescopes to detect gamma rays are difficult to design and operate. Also, it is difficult to separate gamma rays that are cosmic in origin (the ones that interest astronomers) from those that are produced in the earth's upper atmosphere. These events are interesting but they give little information about the galaxy itself, or about conditions beyond the galaxy.

Such are the problems that plague the high-energy astronomer. Nevertheless the work continues. Some have said gamma ray astronomy is really "far out." The statement is true enough no matter how you interpret it. The gamma ray astronomer may be probing into the boundaries of the universe, some 13 billion light years away. Additional information about the sources of gamma rays, their distribution in space, and the mechanisms by which they are produced is being obtained as space-based astronomy develops.

Exploration of high-energy celestial X rays and gamma radiation, and indeed of all parts of the electromagnetic spectrum, is the major concern of astrophysicists. During the last quarter century radioastronomy has become a full-fledged science; so has infrared and ultraviolet astronomy; and now gamma ray and X-ray astronomy is developing. Depending upon wavelength and energy level, sections of the spectrum are being used more and more widely in medicine and chemistry. The frontiers of the spectrum, the ultralong waves and especially the ultrashort waves, are being explored—reducing areas of ignorance.

Hundreds of investigators have studied the electromagnetic spectrum during the past several decades. And new generations of scientists continue to do so. They have probed into the atom, into the microworld; and have surveyed the macroworld of stars and galaxies to determine their structure and function and to bring us closer to understanding what our universe is, and how and when it came to be. They have turned ignorance into understanding, and have shown us that the electromagnetic spectrum is indeed the key to the universe.

FOR FURTHER READING

Asimov, Isaac, *The Universe.* New York: Walker, 1966.

Branley, Franklyn M., *Black Holes, White Dwarfs, and Superstars.* New York: Thomas Y. Crowell, 1976.

———, *Color: From Rainbows to Lasers.* New York: Thomas Y. Crowell, 1978.

Fuller, R. Buckminster, "Heartbeats and Illions." *World,* March 27, 1973, p. 3.

Verschuur, Gerrit L., *The Invisible Universe.* New York: Springer-Verlag, 1974.

Weaver, Kenneth F., "The Incredible Universe." *National Geographic,* May 1974, pp. 589–633.

INDEX

Illustrations are indicated by italics.

aberration of starlight, 18–19, *20*
absorption spectra, *30*, 31–32, 64–65, 75–76, 82–83
Ampère, André Marie, 39
Ångström, Anders Jonas, 48, 65, 69
Arago, Dominique François Jean, 39
atmosphere, shielding by, 2, *2–3*, 52, 88–89, 107, 110, 118
atomic structure, 66–69, *67, 68, 70*, 81–83, 89–90

Balmer, Johann Jakob, 65, 66, 69, 82
batteries, 38–39
Bell, Jocelyn, 95
Bell Telephone Laboratories, 86, 100
Bethe, Hans, 69
Big Bang Theory, 99–100, *99*, 116
black body radiation, 50, *51*, 52, 53
black holes, 119–120, *119*
Bohr, Niels, 47, 55, 67
Bradley, James, 18–19
bright line spectra, 30–31, *30*, 63–64, 65–66, *67*, 75–76, 82–83
Bunsen, Robert Wilhelm von, 75

comets, 62
cosmic rays, 116, 120
Crab Nebula, 96–97, *97*

da Vinci, Leonardo, 23
"deceleration radiation," 117
diffraction gratings, 28–30, 64
"dipping needle," 36
Doppler effect, 76–81

Einstein, Albert, 47, 55, 58, 59, 61, 62, 66, 69
electricity, 5, 11, 15, 24, 36, 38–42, 43–44, 46, 49
  batteries as source of, 38–39
  generation of, 14–15, 24, 38–39, 41, *41*
  light and, 24, 39–42, 43–44, 46, 49
  magnetism related to, 5, 36, 39, 40, 41, *41*

electromagnetic spectrum:
  composition of, 1–3, *2–3*, 4–5, 7, 11–15, *12–13*
  human sensitivity to, 1, 4, 6, 71
  light as clue to, 4, 5, 7–8, 15
  methods of studying, *12–13*
  recent development of, 5–6, 8, 11, 15, 22, 121
  relationships within, 2–3, 4, 5, 8, 11, 14, 42, 46
  wavelengths in, 2–3, 4, 5, 8, *9*, 11–14, *12–13*, 61
electromagnetic waves:
  electromagnetism in, 5, 42, 43–44
  energy in, 7, 10, 11, 14, 49, 50, 52, 53–54, 57, 58–59, 60
  frequency of, 8–10, *12–13*, 14–15
  generation of, 43–45, *44*, 53–54
  light as, 4–5, 8, 15, 23, 24–26, *25*, 41–42, 43–44, 46, *47*
  properties of, 7, 8, *9*, 44, 46
  quantum theory of, 52–55, 59, *68*, 69
  speed of, 4, 7, 8, 42
  temperature and, 48, 49, 50, 51, *51, 54*, 89, 112
  wavelengths of, 2–3, 4, 5, 8, *9*, 10, 14, 60
electromagnets, 39, *40*
electrons:
  early theories about, 66–67
  emission of, 56–59, *57*
  energy levels of, 67–69, *67, 68*, 82–83, 89–90
  excitation of, 31, 66, 67–69, 89–90
  fluorescence and, 59–60
  number of, 69, 81–82
  X-rays and, 60, *61*, 115, 116–117
ether, 22–23, 34
Explorer satellites, 118

Faraday, Michael, 34, 39, 41–42
Fizeau, Armand Hippolyte-Louis, 19–21, 76, 79
fluorescent light, 59–60

Foucault, Jean B.-L., 23
Fraunhofer, Joseph, 64, 69
Fresnel, Augustin Jean, 32, 33, 34
Fuller, R. Buckminster, 6

galaxies:
   distances to, *78*, 79
   Milky Way, (our galaxy), 80–81, 90–92, *91*, 108, 116, *117*
   radiation from, 86, 87, 90–92, *91*, 114, 119
   radio waves from, 86, 87, 90–92, *91*
   red shift in, *78*, 79
Galileo, 17, 37–38, 71–72, 95
Galvani, Luigi, 38, 39
gamma rays, 8, 10, 11, 15, 120–121
Gilbert, Sir William, 36, 37, 39, 43
gravity, 36, 37
Greece, 16, 35, 36

Hahn, Otto, 69
helium, 83
Herschel, Sir William, 48, 73, 91, 101–102, 104
Hertz, Heinrich, 44, 46, 56
Hulst, H. C. van de, 88
Huygens, Christian, 22, 23
hydrogen, 65, 67, *67*, 69, 88, 90, 92

infrared waves:
   characteristics of, 11, 49, 102, *103*, 104, 107
   detection of, 102, 104, 106–107, 108–109
   discovery of, 101–102
   photography with, 104–106, *105*
   stars studied with, 108–109
   Sun as source of, 1, 102
interferometers, 94
Io, 17–18, 95

Jannsen, J., 83
Jansky, Karl, 86, 87
Jupiter, 17–18, 94–95

Kepler, Johannes, 37, 38, 62
Kirchhoff, Gustav Robert, 64, 65, 75

Leeuwenhoek, Antony van, 71
light:
   ancient theories of, 16, 35

light(*con't.*)
   astronomical measurement of, 17–19, 22
   diffraction gratings and, 28–30
   in electromagnetic spectrum, 4–5, 7–8, 11, 15, 23, 34, 39, 42, 43–44, 46, *47*
   ether theory of, 22–23, 34
   interference of, 24–26, *25*
   perception of, 1, 4, 15, 35, 71
   photoelectric effect of, 56–59, *57*
   photons of, 31, 58–59, 62, *68*, 69
   polarized, 32–33, *32*, 41–42
   sources of, 31, 46–47, 66, *67*
   spectrum of, 5, 26–28, *27*, *30*, 48, 71
   speed of, 4, 8, 17–22, 23
   wavelengths of, 4, 8, 26, *27*, 28, 29
   wave nature of, 22, 23–26, *25*, 32, 34, 42, 57, 58, 59
   waves vs. particles of, 22, 23–26, *25*, 28, 29
Lippershey, Hans, 71
lodestones, 35–36
Lyman, Theodore, 66

magnetism:
   earth and, 36
   electricity related to, 5, 36, 39, *40*, 41, *41*
   fields of, 34, 37, 39, *40*, 41, 80, 94
   light and, 39–42, 46, *47*, 80
   lodestones and, 35–36
   poles of, 37
Massachusetts Institute of Technology, 98
Maxwell, James Clerk, 34, 41, 42, 43, 44, 53, 54, 56
Melloni, Macedonio, 102
Melvill, Thomas, 63
Michelson, Albert A., 21, 23
microscopes, 71
Morley, E. W., 23

neon light, 29, 63
neutron stars, 95–97, 114, 118
Newton, Isaac, 22, 23, 24, 26, 27–28, 37, 73
nuclear energy, 69

Oersted, Hans Christian, 39
Orbiting Astronomical Observatories, 112
Orbiting Solar Observatories, 111–112

Parsons, William, 106
Paschen, F., 66, 69
pendulums, 52–53
periodic chart of the elements, 70
photoelectric cells, 49
photoelectric effect, 56–58, 57
photons:
  comets and, 62
  emission of, 31, 68, 69
  infrared, 101, 107
  photoelectric effect and, 58–59
  X-ray, 60–61, 61, 115, 116
Planck, Max, 47, 52, 53, 54, 55, 59, 62, 69
Planck curve, 54, 54, 55
planets, temperature of, 92–95, 108
polarization, 32–33, 32
Polaroid, 32, 33
pulsars, 95–97, 114, 118

quantum theory, 52–55, 59, 68, 69
quasars, 114, 119

radio waves:
  Big Bang and, 99–100
  characteristics of, 10, 11, 46, 84, 88, 89, 95
  discovery of, 8, 45–46, 45
  emission spectra of, 88, 89
  galaxy as source of, 86, 87, 90, 91, 92
  generation of, 45–46, 45, 84
  hydrogen as source of, 88, 90, 92
  interstellar matter and, 97–98
  planets as source of, 92–95
  pulsars found by, 95–97, 97
  radio telescopes and, 84–88, 85, 89, 90, 94
Reber, Grote, 86–87
Roemer, Ole, 17–18, 22
Roentgen, Wilhelm Konrad, 60, 114–115
Rossi, Bruno, 118
Rutherford, Ernest, 47, 55, 67, 69

solar wind, 62
sound, 2, 3–4, 34
spectroheliograph, 111
spectroscopy:
  absorption spectra in, 30, 31–32, 64–65, 75–76, 82–83
  atom studied with, 81–83
  bright line spectra in, 30–31, 30,

bright line spectra in (cont.)
  63–64, 65–66, 67, 75–76, 82–83
  diffraction grating for, 28–30, 64
  Doppler effect and, 77, 78, 79–81, 79
  magnetic fields and, 80
  stars studied with, 75–76, 78, 79–81, 79
  temperature found by, 51, 75
Sun:
  atmosphere of, 64–65, 112, 113
  elements found in, 70, 83, 111, 111
  position of, 80, 91–92
  radiations from, 1, 2, 48–49, 51–52, 54, 62, 64, 69, 87, 110–112, 111
  temperatures of, 51–52, 112, 113
synchrotron emission, 89, 117

telescopes, 19, 26, 28, 71–75, 74, 90
thermocouples, 49, 106–107
thermopile, 107
tourmaline, 33, 42

Uhuru satellite, 118
ultraviolet waves:
  characteristics of, 11, 48, 110, 112
  light and, 1, 60
  photoelectric effect of, 49, 56, 57
  solar temperatures and, 112, 113
  stars studied with, 114
  Sun as source of, 110–112, 111
universe, nature of, 98–99

Van Allen belts, 94
Venus, 93–94
Volta, Allesandro, 38, 39

X rays:
  black holes and, 119–120, 119
  characteristics of, 11, 115, 118
  discovery of, 60–61, 114–115
  electrons and, 60, 61, 115, 116–117
  generation of, 60–61, 61, 115
  natural sources of, 2, 97, 115–117, 118–120
  stars as source of, 116–117, 119
  uses of, 115, 117–118

Young, Thomas, 24–26, 28, 32, 43

Zeeman, Pieter, 80

ABOUT THE AUTHOR

Franklyn M. Branley, Astronomer Emeritus and former Chairman of The American Museum–Hayden Planetarium, is the author of many books, pamphlets, and articles on various aspects of science for readers of all ages. THE ELECTROMAGNETIC SPECTRUM is the ninth in the Exploring Our Universe series, written by Dr. Branley.

Dr. Branley holds degrees from New York University, Columbia University, and the State University of New York at New Paltz. He and his wife live in Woodcliff Lake, New Jersey, and spend their summers at Sag Harbor, New York.

ABOUT THE ILLUSTRATOR

Leonard D. Dank is a graduate of Cornell University and the Massachusetts General Hospital–Harvard Medical School of Medical Illustration. His work has appeared in numerous school texts, health encyclopedias, and children's books, as well as in medical journals and national magazines. Mr. Dank is presently Consultant Director of Medical Illustration at St. Lukes's Hospital Center in New York, and is audiovisual consultant to several major publishing houses. He and his family live in Cutchoque, New York.